Teaching Biology
in Higher Education

Brian Alters
McGill University

Sandra Alters

JOHN WILEY & SONS, INC

To order books or for customer service call 1-800-CALL-WILEY (225-5945).

ISBN 0471-70169-6

Printed in the United States of America

10 9 8 7 6 5 4 3 2 1

Printed and bound by Malloy, Inc.

CONTENTS

CHAPTER 1

Educational Research and Improving Teaching

"Research has taught us a great deal about effective teaching and learning in recent years, and scientists should be no more willing to fly blind in their teaching than they are in scientific research."
Bruce Alberts, President, National Academy of Sciences[1]

THE USEFULNESS OF EDUCATION RESEARCH

Biology instructors often fall prey to the misconception that teaching cannot be improved because teaching is not a science. These instructors generally think that it makes no sense learning so-called effective teaching practices because these practices are almost all non-empirically based. To a large extent, they view good teaching as guesswork or wishful thinking. Are you one of these instructors? If so, would you be surprised to learn that a large proportion of educational research uses scientific methods?

The systematic observation of teachers and students underlies research on teaching and learning, and educational researchers have put a great deal of effort into determining possible cause-and-effect relationships between certain teaching practices and the learning process. Scientific methodologies have contributed to educational research more than most biology instructors are aware. For example, statistical methods of analysis of variance and path analysis, developed by evolutionary biologists and used to characterize and analyze variation, are commonly used in quantitative education research. Investigators in many fields and over many decades have conducted a significant amount of higher education research that biologists would find methodologically sound, testable, and verifiable.

Research in education often has methodological difficulties, however, and these difficulties may be the root of some misconceptions that exist as to its rigor (or lack of rigor). One of the primary methodological difficulties lies in the concept of controls. In many biological experiments, all but one or a few variables often can be controlled, and the researcher can therefore attribute possible effects to the experimental treatment. In educational research, control of variables is difficult and often impossible because students undergoing the experimental treatment (e.g., a new teaching technique) have a variety of differing experiences during the treatment period. For example, if the experimental duration is one month, with four one-hour class sessions of treatment with pre- and post-testing at the beginning and end of the experimental month respectively, then the students have hundreds of hours of experiences that may affect the experiment and that are not part of the treatment. More specifically, the treatment may have been a new teaching technique for introductory evolutionary concepts and, outside of class, some students may have seen an educational TV program that discussed evolution, attended another course that touches on evolution, read about evolution, discussed evolution concepts with knowledgeable siblings or roommates, and so forth. Naturally, the more specialized and specific the treatment, the lower the probability that outside factors such as television, other courses, additional reading, and discussions with siblings would affect the experiment.

To help alleviate the concern about appropriate controls, some educational researchers conducting quantitative experimental designs use control groups with the assumption that students in the control group and the treatment group will have essentially the same experiences outside of the classroom. Therefore, any differences between the control group and the treatment group can likely be attributed to the treatment. Educational researchers who use control groups often prefer to randomly assign students to each group to lessen sampling error. Such methodology may seem extremely straightforward, but randomly assigning students to classes for treatment is difficult since students generally select their own classes, and student class selections may be based on variables that could affect the experimental outcome.

In addition to possible problems with assuring appropriate controls, other threats to validity of quantitative designs in educational research exist and are due to the nature of working with human subjects. One complication is that research with human subjects often necessitates working within human subject committee regulations. (Biologists who find animal ethics committee regulations as somewhat inconvenient or time-consuming can probably appreciate the complicating aspects of human subject committee regulations.) In addition, the understandings, intentions, and values of human subjects can affect bias, implementation, and funding. And even when experimental design results/conclusions are beneficial and are then used in larger arenas, rarely does the treatment have just one significant effect. Yet despite difficulties with controls, other threats to validity, and the problems inherent with working with human subjects, educational research is often rigorously experimental and can provide trustworthy scientific evidence on which decisions about change for improving instruction can be based.

Maybe you are still not convinced of the efficacy of educational research. You may feel justified in thinking that educational research is not as exact a science as is the stereotypic natural science experimentation models. In addition, you may agree with the comments of University of Maryland theoretical physicist Edward F. Redish when he notes in *Teaching Physics with the Physics Suite*, that "...neither the educational phenomenology growing out of observations of student behavior nor the cognitive science growing out of observations of individual responses in highly controlled (and sometimes contrived) experiments has led to a single consistent theoretical framework. Indeed, it is sometimes hard to know what to infer from some particular detailed experimental results."[2] In defense of educational research, however, Redish notes that it deals with an extremely complex system, and he continues to explain that educational research results have revealed what does *not* work in the classroom. He suggests that educational research results are useful to help instructors focus on their students and not just on the content to be taught. We suggest that at the very least, college and university biology instructors consider giving educational research "a chance" to determine its usefulness in improving their biology teaching.

Many suggestions are given in this book, and probably no one would, or could, implement them all. Some depend on funding, scheduling, venue, and, most importantly, time. Most can be implemented fairly easily. However, implementing only some of the suggestions should make a difference in your teaching. Choose suggestions to try based on what you think is most doable in your classroom. You might think of the suggestions you choose as "treatments" in experiments you conduct concerning your teaching. Virtually all the suggestions we cite in this book are based on the research reports of university researchers. Sometimes, as with most research, there are contradictory conclusions by researchers. Even in those cases in which contradictions exist, instructors can "run the experiment" themselves to see what occurs in their course contexts.

CONDUCTING, PROMOTING, AND IMPLEMENTING EDUCATIONAL RESEARCH

Many professors and other education professionals conduct research in education. For example, more than 20,000 educators, administrators, directors of research, counselors, evaluators, graduate students, and behavioral scientists are members of the American Educational Research Association (AERA). The AERA encourages scholarly inquiry related to education and promotes the dissemination and practical application of research results.

Although the AERA is the most prominent international professional organization with the primary goal of advancing educational research and its practical application, a variety of other organizations promote educational research as well, including The Society for Research into Higher Education (SRHE). In addition, the National Science Foundation (NSF), the American Association for the Advancement of Science (AAAS), and the National Research Council (NRC, a division of the National Academy of Sciences [NAS]) all support research in biology education.

In 2002 the Committee on Scientific Principles for Education Research of the National Research Council produced a book-length report *Scientific Research in Education* that provides information on evidence-based educational research. The charge of the NRC committee was "to examine and clarify the nature of scientific inquiry in education." In addition, they were to ". . . review and synthesize recent literature on the science and practice of scientific educational research . . ."[3] The report supplies six guiding principles that not only underlie educational research but also all scientific inquiry:

1. pose significant questions that can be investigated empirically
2. link research to relevant theory
3. use methods that permit direct investigation of the question
4. provide a coherent and explicit chain of reasoning
5. replicate and generalize across studies
6. disclose research to encourage professional scrutiny and critique[4]

Although research in science education and in education in general is widely supported and conducted, implementation of results is not as widespread. On this point, the NRC report notes:

> *. . .the generation of scientific knowledge—particularly in social realms—does not guarantee its public adoption. Rather, scientific findings interact with differing views in practical and political arenas . . . . The scientist*

discovers the basis for what is possible. The practitioner, parent, or policymaker, in turn, has to consider what is practical, affordable, desirable, and credible. While we argue that a failure to differentiate between scientific and political debate has hindered the scientific progress and use, scientific work in the social realm—to a much greater extent than in physics or biology—will always take place in the context of, and be influenced by, social trends, beliefs, and norms.[5]

THE LINKS BETWEEN SCIENCE RESEARCH AND SCIENCE TEACHING

Professors at research institutions often complain that they have little to no time to contemplate how they might improve their teaching. Some professors adhere to the unwritten rule of spending no more than one hour of preparation time for one hour of class time. Such a rule has no basis in empirical evidence for producing good teaching, but adherence to "the rule" exists in many biology departments that clearly value empirical evidence in their scientific work. Why? One main reason is that professors at research institutions perceive that research is valued more highly than teaching for promotion and tenure. Virtually all professors at research universities know that leading researchers can be terrible instructors and still be tenured and promoted. Nevertheless, there does appear to be a slight shift taking place at research institutions, and quality teaching is becoming valued more highly than in the past for promotion and tenure. A biology higher education research report from the NRC recognizes the need for quality teaching to be valued and notes that:

"Departments and colleges must find new ways to help individual faculty and academic departments innovate and reward their efforts in creating, assessing, and sustaining new educational programs. For example, faculty interested in adapting teaching approaches for their own use or in creating new teaching material should have lighter than normal requirements for teaching, research, or service while actively engaged in such projects. Also, travel funds earmarked especially for faculty development or education meetings should be provided to enable faculty to participate in meetings that enhance their teaching capabilities."[6]

Meta-analyses of research results on the correlation between research productivity and instructional effectiveness tend to find little-to-no relationship between the two.[7] Nevertheless, many professors suggest that their research productivity, research skills, and resultant depth of content knowledge enhance their teaching effectiveness. How? Some say that they can teach about cutting-edge research in the field even in large undergraduate classes. Other research professors contend that their reading for research purposes keeps them more up-to-date for their classes than will reading, for example, the textbooks and ancillaries. Those who believe that these research/teaching links are weak contend that only a few class sessions—at most—could be devoted to the professor's research. In addition, for most lower-division courses, much of what is taught is so fundamental that major changes don't occur often. Textbooks for those courses are typically revised on three-year cycles and include changes in the field. In addition, most biology instructors become informed of major changes in their field whether or not they conduct scientific research.

But there are other research/teaching links. Understanding biology involves understanding how biologists do their work. Many advocate that research is better illustrated for students in courses where an interdisciplinary nature of biological research is explicitly one of the points of the course or, better yet, *is* one of the courses. The NRC report *Bio 2010: Transforming Undergraduate Education for Future Research Biologists* recommends that "In addition to modules, interdisciplinary lecture and seminar courses can give students a better and more realistic picture of how connections between different areas of science are made in research. Because research is becoming increasingly interdisciplinary, such courses should be made available to students beginning in their first year."[8] This report also recommends involving undergraduates in their own research or in faculty research, and states that "All students should be encouraged to pursue independent research as early as is practical in their education. They should be able to receive academic credit for independent research done in collaboration with faculty or with off-campus researchers."[9] Graduate and/or postdoctoral students can help direct undergraduates' research for professors who do not have the requisite time or who have large classes. Undergraduates may also become involved in such activities as presenting poster sessions at academic meetings.

Another way to link research and teaching is via seminar courses, which can educate and stimulate undergraduate students throughout their undergraduate years. Many undergraduates become more engaged and excited in seminar courses than in traditional courses because of the discussion format, which generally involves "real" problems and "real" research. This excitement not only generates enthusiasm for research but also for learning because students begin to make connections between class work, research, and problem-solving. Moreover, seminar courses are useful in institutions in which it is difficult to have undergraduates conduct research. In seminar, students can at least read, examine, and discuss biological research. Institutions that do have research opportunities available for freshman and sophomores may benefit students by offering seminar courses in the first year of undergraduate studies to give undergraduates closer contact to research before their junior year. The NRC report *Bio 2010* recommends seminars to communicate the excitement of biology:

> "*Seminar-type courses that highlight cutting-edge developments in biology should be provided on a continual and regular basis throughout the four-year undergraduate education of students. Communicating the excitement of biological research is crucial to attracting, retaining, and sustaining a greater diversity of students to the field. These courses would combine presentations by faculty with student projects on research topics.*[10]

In summary, there are many linkages between research and teaching, and these linkages can certainly enhance teaching. Whether or not you are a researcher, consider bringing research into your teaching. Jenkins, Breen, and Lindsay note in *Reshaping Teaching in Higher Education* that, from the perspective of student learning, "linking teaching and research is achieved when:

- Students learn how research within their discipline leads to knowledge creation.
- Students are introduced to current research in their disciplines.
- Students learn the methods used to carry out research in the disciplines.
- Students are motivated to learn through knowledge of and direct involvement in research.
- Students carry out research.
- Students participate in research conducted by their lecturers.
- Students learn and are assessed by methods resembling research procedures in their discipline.
- Students learn how research is organized and funded.
- Students become members of a school or department and university culture within which learning, research and scholarship are integrated.
- Students' learning is supported by systems and structures at departmental, institutional, and national level that facilitate staff scholarship and research in the pedagogy of the disciplines as well as the disciplinary scholarship and research."[11]

SOME PRINCIPLES OF GOOD TEACHING[12]

All it takes is reading a few books and research articles about teaching in higher education and one will come to a quick realization that there is no one "best way" to teach biology or any other discipline. Nevertheless, the research literature on college science teaching suggests that students learn in ways that mirror research processes—with direct experience and in a manner that is inquiry-based. Majors and nonmajors alike benefit from science teaching that:

- is investigative;
- is often collaborative;
- comes from working on complex, often real world problems;
- engages students in interpreting data and where possible, in gathering their own data;
- shows students the limitations as well as the powers of particular scientific ways of knowing.[13]

In addition, the following are properties of good teaching as seen from the individual instructor's point of view:

- A desire to share your love of the subject with students;
- An ability to make the material being taught stimulating and interesting;
- Facility for engaging with students at their level of understanding;
- A capacity to explain the material plainly;
- Commitment to making it absolutely clear what has to be understood, at what level, and why;
- Showing concern and respect for students;
- Commitment to encouraging student independence;
- An ability to improvise and adapt to new demands;
- Using teaching methods and academic tasks that require students to learn actively, responsibly, and cooperatively;
- Using valid assessment methods;
- A focus on key concepts, and students' misunderstandings of them, rather than on covering the ground;
- Giving the highest quality feedback on student work;
- A desire to learn from students and other sources about the effects of teaching and how it can be improved.[14]

Lectures are, of course, a part of teaching biology in higher education. Teaching facts, generalizations, principles, or concepts – making a variety of points – are key to the lecture format. The following techniques can help you teach a point in lectures:

- Dynamically using gestures, eye contact, vocal inflections, and unscripted speech;
- Exhibiting enthusiasm;
- Keeping statements concise;
- Displaying key words, phrases, or formula/equations;
- Explaining the same thing in different ways;
- Elaborating when necessary;
- Increasing detail when students can handle it;
- Using appropriate illustrations;
- Supplying helpful analogies;
- Employing asides relevant to the lecture goals;
- Providing context for the content;
- Preplanning good examples;
- Ending with a well-crafted recapitulation;
- Proceeding at the pace of the class;
- Using definitions that are clear to students.[15]

Teaching and teaching methods are described in more detail in Chapter 2.

THE PEDAGOGICAL USEFULNESS OF FOSTERING STUDENT INTEREST IN BIOLOGY

Instructors know that most students take introductory nonmajor's biology to satisfy the science requirement for their major outside of biology, and instructors assume that these students are generally uninterested in the subject. Conversely, instructors often assume that those majoring or minoring in biology have a great deal of interest in biology. Regardless of whether the students are majoring or minoring in biology, or are nonmajors, they all likely vary in their degree of interest in the subject matter, and all will benefit from having increased interest. Research results and student evaluations indicate that increased student interest in the subject is almost always beneficial to learning in general, and as student interest increases, long-term retention of concepts increases. In addition, as student interest increases, student pleasure in learning increases. If students enjoy what they are doing, be it academics or not, then they generally work harder and longer. Students often need and enjoy being taught in a manner that increases their interest in what they are currently studying. Even though we may think that cell structure, protein synthesis, or animal behavior is inherently interesting, many students do not.

Offering interesting and exciting courses may also help increase the number of students who elect biology as a major, and this increase can affect numbers of students entering careers in research. *Bio 2010* concluded, among other things, that:

> *"Offering exciting introductory courses will help attract more students to enroll in biology courses, increasing the number who might consider biomedical research as a career. Increasing the number of students who consider biology as a major may increase the quality of future biomedical researchers."*[16]

Why aren't more instructors making their classes more interesting to students? There are many reasons, but research tell us that one primary reason is that biology instructors teach in approximately the same manner in which they were taught when they were students of biology. Copycat teaching may not model the worst biology instructors, but even if it models the best instructors, it is likely to leave opportunities for improving student interest. Why do we assume that the best instructor, whom we think interested us the most, modeled the best ways of increasing interest in a wide variety of students?

What are some ways to increase student interest in a biology course? One strategy is to use focus groups and/or design anonymous surveys asking students what interests them most within biology. Use their suggestions as best you can in the design of the course. Another way to increase student interest is to place biological content within a context, such as telling the story of a discovery rather than just stating the fact that something was discovered. Not only do stories raise student interest levels but, for many students, they also make the content more memorable because it is placed within a larger, everyday, "real life" context. A third way of increasing student interest is to demonstrate how a specific biological method, theory, or research result is applicable to their lives. Even if the content is historical, try to show how knowing about the past helps direct what we do today and in the future, such as eliminating methods that have been tried repeatedly in the past and failed.

Some instructors contend that to make biology more interesting to students, they must "dumb it down." This is untrue; there are many examples of biology instructors that designed courses and individual classes to greatly interest students without sacrificing academic rigor.

STUDENTS, FEAR, AND BIOLOGY COURSES

Many research studies clearly show that yes, many students are afraid of taking a biology course. They are afraid of embarrassment in the classroom or labs in front of their peers, teaching assistants, and/or instructors. They are afraid of failure because many students have little science background, are unable to think critically, and find science concepts difficult to understand and retain. Many non-science majors are quick to point out that they are not good at science, and they may have a negative or indifferent attitude toward science. In addition, they may lack self-discipline and study skills, and may have had negative experiences previously. And now they are forced to take biology. Biology majors or minors may be afraid of their biology courses because poor performance could result in their not being able to continue in their chosen course of study. Although the mean age of students in introductory classes has risen, even many older and more mature students feel somewhat fearful or uneasy in biology class. To quell this fear, one veteran biology instructor suggests:

> *Few students learn about science the way you do, are as interested in science as you, or have the background that you have (or expect them to have). The key then is not to teach the course as if you are talking to yourself or your colleagues. You must identify what problems you're students possess, and then adapt your methods of instruction to meet their needs, experiences, and interests. You should think about how you will promote self-discipline and learning skills in your students, how to broaden student perceptions of and improve student attitudes towards biology, how to improve your students' thinking skills, and how to organize your course so that students leave with a solid background in the major biological concepts. Rule of thumb: don't assume your students already understand or are familiar with any topic or term. Introduce each topic with basic information or define each term, check to see if most students understand, and, if so, move along quickly to the next level.*[17]

Instructors can help students feel more comfortable in biology class, but many instructors have a need, think it is appropriate, and/or believe it is pedagogically sound to instill anxiety and fear in students. Some instructors attempt to make the material seem more difficult than it really is, rather than making the students feel that they can master it. Fear may be a good motivator for some things, but research studies reveal that academic achievement is rarely helped by student fear. Fear may motivate students to spend more time reading and studying, but fearful students often have trouble concentrating when they study, focusing their attention on lectures, and reasoning and thinking in labs.

Instilling anxiety in students also thwarts their risk-taking. Students must feel at ease in order to take risks. While the instructor may not think that asking questions, showing ignorance, or stating things incorrectly is risky, multitudes of students certainly do. You might want to try this simple experiment to convince yourself of the anxiety-risk taking connection. Ask a multiple-choice question about a topic recently discussed in class. Depending on the fear factor involved in the class and the particular question, probably 10–60% of hands will go up. Then ask the same question with the anonymity of remote transmitters (personal response systems). The response rate will increase dramatically. In our classes, responses increase by 30 to 40% when using transmitters and this increase occurs in courses in which the instructors try their best to reduce student unease.

Instructors can reduce student fear or anxiety in biology class by modifying their attitude and behavior. Show students respect, courtesy, and compassion. Try to instill in them a "can do" attitude. This does not mean coddling them, inflating their grades, or passing students who fail. It also does not mean giving in to student requests for more points when they are not deserved. It does not mean sacrificing the rules that the instructor, department, or institution has put into place. Course standards are to be upheld, point systems adhered to, and dates for examinations and lab write-ups kept as per schedule. The idea is to put students at ease within the rules and rigor of the course, which should contribute to more effective teaching. Projected arrogance, uncaring, inappropriate rigidity, and unnecessary stringency and toughness will detract from the goal of increasing students' understanding of biology.

THE LINKS BETWEEN BIOLOGY CURRICULA AND BIOLOGY TEACHING

The biology curriculum is an important part of the teaching/learning equation. The NRC report *Bio 2010* makes numerous recommendations concerning ways to enhance undergraduate education in biology via curricular change. While the chapter on biology curriculum change should be read in its entirety because of various useful details, explanations, and caveats, the following are the concise recommendations made in the report.

> *"Recommendation #1: given the profound changes in the nature of biology and how biological research is performed and communicated, each institution of higher education should reexamine its current courses and teaching approaches . . . to see if they meet the needs of today's undergraduate biology students. Those selecting the new approaches should consider the importance of building a strong foundation in mathematics, physical, and information sciences to prepare students for research that is increasingly interdisciplinary in character. The implementation of new approaches should be accompanied by a parallel process of assessment, to verify that progress is being made toward the institutional goal of student learning.*
>
> *Recommendation #1.1: Understanding the unity and diversity of life requires mastery of a set of fundamental concepts. This understanding will be greatly enhanced if biology courses build on material begun in other science courses to expose students to the ideas of modeling and analyzing biological and other systems.*
>
> *Recommendation #1.2: Biology majors [should] receive a thorough education in chemistry, including general chemistry and aspects of organic chemistry, physical chemistry, analytical chemistry, and biochemistry, incorporated into a new course or courses. . . . Biology faculty should work in concert with their chemistry colleagues to help design chemistry curricula that will not only foster growth of aspiring chemists, but also stimulate biology majors as well as students majoring in other disciplines. Furthermore, chemistry faculty must work with biologist to find ways to collaborate on the incorporation of chemistry concepts, and those of other scientific disciplines, into their teaching of biology. Learning biology should not be dependent upon chemistry but, rather, integrated with it. Biology students should begin their study of chemistry in the first year so that they will acquire a strong foundation in chemistry before starting their study of chemically based aspects of biology.*
>
> *Recommendation #1.4: Life science majors [should] be exposed to engineering principles and analysis. This does not necessarily require that they take a course in a school of engineering; courses in physics, biology, and other departments can provide exposure to these concepts. Students should have the opportunity to participate in laboratories that give them hands-on experience, so that they may learn about the functioning*

of complex systems, especially as they relate to the basic principles of physical science, mathematics, and modeling. Basic courses in physics and engineering should be developed specifically for life sciences students; these courses could be taught in either the physics or the biology department. This could be complemented exceptionally well by biology lecture and laboratory courses that assist students in their understanding of principles of physics and engineering (e.g., a unit on biomechanics taught in a physiology or anatomy course).

Recommendation #1.5: Quantitative analysis, modeling, and prediction play increasingly significant day-to-day roles in today's biomedical research. To prepare ... biology majors headed for research careers need to be educated in a more quantitative manner, and such quantitative education may require the development of new types of courses. . . . Life science majors [should] become sufficiently familiar with the elements of programming to carry out simulations of physiological, ecological, and evolutionary processes. They should be adept at using computers to acquire and process data, carry out statistical characterization of the data and perform statistical tests, and graphically display data in a variety of representations. Furthermore, student should also become skilled at using the Internet to carry out literature searches, locate published articles, and access major databases.

Recommendation #2: Concepts, examples, and techniques from mathematics, and the physical and information sciences should be included in biology courses, and biological concepts and examples should be included in other science courses. Faculty in biology, mathematics, and physical sciences must work collaboratively to find ways of integrating mathematics and physical sciences into life science courses as well as providing avenues for incorporating life science examples that reflect the emerging nature of the discipline into courses taught in mathematics and physical sciences.[18]

As we discussed previously, there are many linkages between research and teaching. So how do you know when appropriate links have been made? Instructor and student perspectives may differ greatly. From an instructor perspective, gains have been made in linking the curriculum and research when the curriculum incorporates more: biology research developments, learning experiences embedded within a context of research, and requirements for students to conduct research projects on their own or with staff researchers. From a student perspective, appropriate curricular-research links have been made when students: are assessed by methods similar to those used in their research (sometimes referred to as "authentic assessment"), feel that they can conduct the bulk of their research on their own, can explain the research they do to non-biologists, feel comfortable with the research, can teach it to lower-level classmates, and are motivated to do more research.

There are many other factors to take into account when developing a curriculum that does a good job of linking research and teaching. These factors include what the students want, what the administration wants, what instructors desire students to know and perform, what students learn inside and outside of class time, what technologies will be employed, what learning methods will be employed, what biological and educational research literature will be consulted, what concepts of the discipline should be taught, what time scheduling and credit distribution changes should occur, and what effect implementing the new curriculum has on the budget. Changing curricula to better link teaching and research is more than just a matter of simply increasing the mention of research in lectures or labs. In addition to research as a curricular theme, some other useful curricular themes are nature of science, evolution, technology, personal aspects of biology, social aspects of biology, ethical aspects of biology, and the historical development of biological ideas.

For nonmajors' introductory biology, the research literature suggests a variety of goals that are useful to address when designing the curriculum for this course. At the conclusion of an introductory biology course, nonmajors should be able to

- appreciate and value science;
- make informed decisions about science-related personal and social issues;
- read, understand, critique, and discuss popular scientific information;
- defend positions on science-related issues;
- apply scientific processes of investigation to daily life;
- make sense of the natural world;
- understand the major principles of biology and science as a process; and
- understand the role of humans in the biosphere.[19]

More generally, the Society for College Science Teachers' position statement on introductory college-level science courses states that, "at a minimum, every student should know and be able to do the following:. . . Solve and evaluate problems. . . . [Design] meaningful experiments. . . . Evaluate critically. . . . Explain scientifically related knowledge claims. . . . Ask meaningful questions about real-world scientific issues."[20]

CONDUCTING EDUCATIONAL RESEARCH IN A BIOLOGY DEPARTMENT

Should scholarship of teaching and learning be regarded as scholarship within a biology department? Should biology faculty members approach their teaching in a more scholarly manner? Should they treat their teaching as experiments not only to improve their own teaching for their students but also as a serious form of scholarship similar to their biological scholarship? Consider the career of a distinguished biologist who is also a distinguished researcher in higher education.

Craig E. Nelson has been a research biologist and biology instructor for almost 40 years at research-intensive Indiana University (1966–2004, now Emeritus). Nelson started his efforts and scholarship of teaching and learning decades ago and progressed to being awarded the highest honor bestowed by Indiana University in 2001, the *President's Medal for Excellence*. In addition, he was awarded the Carnegie Foundation's *Outstanding Research and Doctoral University Professor of the Year* (2000) for the United States, and he has received numerous other teaching and research awards over multiple decades.

Nelson feels he knew nothing of the scholarship of pedagogy when he began teaching at Indiana University, and that his views of teaching were very conventional. (He and other biology instructors were not exposed to pedagogical scholarship in graduate school.) He saw that course content had been updated over the decades while pedagogy changed very little. He thought his teaching career would reflect these typical realities and saw nothing particularly detrimental that would encourage him to change his teaching. Only a few years later Nelson started to change his teaching goals because, as Nelson states, "I was no longer satisfied with students just learning the science and then ignoring or forgetting it."[21]

The more he began reading for teaching, the more he changed his research and refined his teaching goals. His teaching goals began to have less emphasis on content and more on critical thinking. What he wanted to accomplish in teaching was to improve student thinking and decision-making skills as well as increasing content understanding. He reframed how he thought about teaching and learning by reading numerous theoretical works and empirical studies, and began researching teaching and learning in higher education.

Nelson successfully argued at Indiana University that teaching and learning scholarship should be officially regarded as scholarship within the biology department, and decided to conduct both biology and biology education research. His beginning efforts in higher education research required his mastering the literature, learning a new technical language, writing educational grants (with subsequent awards), and publishing in this new field of scholarship. With regard to Nelson's publications he writes, "these publications were not simply a reporting of teaching experience, but rather required a synthesis of my own experience with diverse theoretical and empirical studies of pedagogy."[22]

As in science, he found that much of educational research is evidence-based and theory-framed. He learned that structured student-student interactions in collaborative learning is extremely important in advancing critical thinking skills. He began teaching a graduate course on pedagogy in which ideas that emerged from the course would be used in teaching his undergraduate biology courses.

The incorporation of biology education concerning teaching and learning in biology departments is not new. As far back as 1975, Nelson's department had a course for Ph.D. students on teaching college biology, which he has taught at least a dozen times since that date. Some of the key topics he addressed in that course included student difficulties in understanding scientific ideas, alternative frameworks for developing critical thinking, learning styles and their use in designing instruction, problems of content coverage versus depth, and traditional grading approaches. Today, numerous large and small institutions offer courses for their Ph.D. students on the research and practice of undergraduate teaching and learning.

Traditional biology instruction can be changed and improved. Professor Nelson found ample reason why typical instruction in higher education should change, and he conducted research to that end in a biology department. As a consequence, Nelson greatly improved his own biology teaching and has helped others improve as well.

CHAPTER 2

Teaching & Learning

"If we cannot help students to enjoy learning their subjects, however hard they may be, we have not understood anything about teaching at all."
Paul Ramsden, Pro-Vice Chancellor of Teaching and Learning, University of Sydney[1]

Defining "good teaching" is difficult; many academics have suggested various definitions. One rather concise and practical definition, suggested by John Biggs in *Teaching for Quality Learning at University,* is "Good teaching is getting the most students to use the higher cognitive level processes that the more academic students use spontaneously."[2]

DEVELOPING CRITICAL THINKING SKILLS

What are higher cognitive level processes? Biggs suggests the following continuum of cognitive processes, with concomitant levels of student engagement. Beginning with the lowest, these levels are memorizing, notetaking, recognizing, relating, applying, generating, reflecting, and theorizing.[3] Benjamin Bloom's Taxonomy also provides a similar continuum of cognitive function.[4] This guide has been well-known for decades and is useful for classifying cognitive types of questions. These levels of knowledge, from lowest to highest, include knowledge, comprehension, application, analysis, synthesis, and evaluation. The levels of application and above are considered higher-order thinking or critical thinking. Each level in this taxonomy subsumes the levels below.

While some students are skilled in critical thinking processes, other students are not and take considerably more time with such cognitive tasks. One objective of teaching is to help students develop critical thinking skills. In addition to helping students hone their critical thinking abilities, many biology educators consider the following student outcomes, in the areas of skills, knowledge, and values, as important. Consider helping students develop the following skills: critical evaluation, communication, manipulation, computation, observation, and application. In addition, consider helping students become knowledgeable about the nature of science, evolution and integrating principles, biological concepts, personal/social/ethical aspects of biology, the nature of technology, and the history of biology. Values that might be promoted include curiosity, openness to new ideas, skepticism, a respect for logical and empirical approaches, integrity, diligence, fairness, imagination, and consideration of consequences.[5]

How can an instructor help develop critical thinking skills in students, along with the skills, knowledge, and values we just listed? There are many teaching models to choose from that are based on ideas of how students learn. Virtually every method has its supporters and critics. We will present some of the more popular methods. You can combine methods, or you can use different methods to teach different concepts.

As most instructors know, students not only have problems retaining what they "learned" in biology, but they also have problems retaining in other disciplines. In fact, most students retain 20% or less from their courses. The National Science Board in *Science and Engineering Indicators 2000* reported that only about one-third of college graduates (baccalaureate) could accurately answer questions such as "What is a molecule?" and "What is the Internet?" In addition, about one-third of graduates thought that lasers work by focusing sound waves and did not know how long it takes for the Earth to go around the sun. Less than half knew that the universe began with a huge explosion.[6] Educational documentary films by the Harvard-Smithsonian Center for Astrophysics have shown the inability of graduating Harvard seniors to explain why it is hot in the summer and cold in the winter, and the extreme difficulties of MIT graduates to complete a simple DC circuit when given a flashlight battery, lightbulb, and a piece of wire.[7] These basic concepts are all used as a foundation on which to build more complex concepts. What happens to student understanding of more complex concepts when the fundamental concepts on which they are built are misunderstood?

Certainly, instructors teach the fundamentals of science; so why are students not learning what is being taught? Depending on the concept, students "learn" superficially for the short-term and do well on course examinations but given time it is evident that numerous students have not changed their prior conceptions. When asked the same questions again at a later time, the students revert to the incorrect conception they held prior to the instruction. Extinguishing students' prior conceptions is a difficult process that may require particular teaching methods.

The job of the instructor is to help students construct meaning from their learning experiences. Only the students can do it; instructors cannot "fill" them with knowledge. Many instructors admit that they simply "present the material," attempting to transfer or download information from teacher to student, and the students must compete in a sink-or-swim selection process. Certainly students will rise to the top levels of their class and do well on examinations, but that is not the point. The point is that even Harvard and MIT graduates who have done precisely that, often appear to have trouble recalling fundamental scientific concepts that they supposedly learned to become the top in their classes. The scientific literacy among many college graduates – including many "smart" college graduates—is shockingly limited.

Traditional methods may not the best to promote student learning. In a review of three decades of research literature on higher education, Rutgers University zoologist Lion F. Gardiner[8] summarized important aspects of what hampers student learning with regard to traditional lectures, critical thinking, testing, and curriculum:

Lectures. Many students are strikingly limited in their ability to reason with abstractions. Therefore, they have significant difficulty understanding many college lectures that instructors perceive to be straightforward and level-appropriate. In addition, student attention begins to drift only 10 to 20 minutes into a lecture. Students are rarely involved in frequent student-instructor or student-student instructional interaction during class; in many classes students are asked questions and respond to them for less than 10% of the class time. This paucity of interactive learning is especially noteworthy given that there is an inverse relationship between lecture listening time and critical thinking. In addition, there is a positive association between lecture listening time and rote memorization. Even when professors do ask questions of students in class, approximately 90% of the questions are the knowledge-based recall type, with just a few percent requiring higher order (critical) thinking skills.

Critical thinking. Most professors want students to be critical thinkers. However, many students view the academic world in terms of true or false, right or wrong, and credit or no-credit. Instead of analyzing evidence that contradicts their erroneous conceptions, students often just passively receive knowledge from authorities – professors – and attempt to memorize it for the test. To become active learners, students need an environment in which learning is a social process in which they are taught and can practice cognitive skills such as analysis, synthesis, and evaluation. Nevertheless, only 10% to 30% of professors use methods other than traditional lectures as their primary pedagogy.

Testing. The attempt to engender higher order thinking skills in students is often thwarted by the very way instructors evaluate student work. The knowledge-level tests that are typically administered in university science classes reinforce the "right or wrong" mentality, concrete thinking, and mere surface approaches to learning. Additionally, the validity and reliability of many such tests can be called into question. Moreover, assessment is often infrequent, which diminishes effective educational experiences.

Multiple choice tests, which are often a necessity in large lecture hall classes, can be designed to be reliable and powerful assessment tools. To accomplish this task, the distractor options (wrong answers) should closely match student misconceptions. The use of such questions as part of an interactive teaching approach has been quite important in demonstrating the relative ineffectiveness of traditional pedagogy in college physics.[9] (See Sadler for a research-based explanation of developing distractor-driven assessment instruments.[10])

Curriculum. Although important, the content and structure of the curriculum is not as important to learning as how a professor structures the learning environment and how students approach the curriculum.

CONSTRUCTIVISM

In the last couple of decades, cognitive psychologists and science educators have proposed a model for teaching based on learning theory called *constructivism.* From a constructivist perspective, learning is a social process in which students make sense of experience in terms of what they already know.[11] Instructors who use a constructivist-based teaching approach provide students with situations to examine the adequacy of their prior conceptions. Facilitating instruction from a constructivist perspective, therefore, is totally unlike traditional college teaching, in which instructors simply tell or explain to students what they should know. In a constructivist approach, the instructor uncovers misconceptions the students hold and designs opportunities for the students to see, as directly as possible, that their ideas are somehow inaccurate. The instructor does not *tell* the students that their ideas are inaccurate, but they arrange activities that result in students coming to that conclusion on their own.

There are a variety of constructivist theories and related approaches. Cognitive scientist and biology educator Anton E. Lawson of Arizona State Uni-

(1) Present a set of initial parameters (fitness of both homozygotes equal that of the heterozygote, $p = 0.3$ $q = 0.7$, large population). Ask the students to predict: "What will happen through time? Will there be an equilibrium or fixed end point? If so at what frequency?" Then ask them to graph their predictions (as allelic frequency vs. time). In this case, some usually will predict that the "dominant" (AA) will quickly prevail—because it is "dominant" in the vernacular sense. (2 & 3) Present a computer simulation that graphs an allelic frequency versus time for several populations (or, less excitingly, present the graph as an overhead). When a series of conditions and predictions are examined, the simulations will contradict some of the predictions for many groups of students. This allows them to find their prior misconceptions and revise their mental models. (4) After each round, the new ideas can be tested against new sets of conditions embodying the same genetic concepts until the predictions match the new (subsequently produced) results.[14]

Box 1 Population Genetics Example

versity has distilled some important aspects of constructivism into *essential elements of constructivist instruction*:

1. Questions should be raised or problems should be posed that require students to act on the basis of prior beliefs (concepts and conceptual systems) or prior procedures.
2. Those actions should lead to results that are ambiguous or can be challenged or contradicted. This forces students to reflect back on the prior beliefs or procedures used to generate the results.
3. Alternative beliefs or more effective procedures should be proposed by students and the teacher.
4. Alternative beliefs or the more effective procedures should now be utilized to generate new predictions or new data to allow either the change of old beliefs or the acquisition of a new belief (concept).[12]

A constructivist approach helps students see that science is not just a set of conclusions to be memorized but is a process of proposing alternatives to be tested. Box 1 gives an example of a constructivist approach to teaching population genetics. This approach has a positive effect on student critical thinking abilities because students are asked higher order thinking questions and are directed to think through alternative conceptions. And critical thinking/reasoning is linked to scientific beliefs. Lawson and Weser found a negative correlation between reasoning abilities in nonmajor introductory biology students and their holding nonscientific beliefs.[13]

Students *can* learn in a constructivist manner when they are "told things" by their instructors, as they attempt to construct new knowledge from what they hear and what they already know. However, this "telling" method alone is non-constructivist-based teaching. Students in this type of teaching may attempt to actively construct new knowledge when they hear the lecture. However, this "telling" method alone is non-constructivist-based *teaching*. There is sufficient evidence that learning is enhanced when students are taught via constructivist-based teaching methods as discussed in the preceding paragraphs.

One basic difference between such "active" teaching methods and traditional college-level instruction is student-student discussion. Numerous research studies indicate that increasing student-centered discussions increases retention, problem-solving ability, and knowledge transfer to other situations.[15] Some physics education research has suggested such methods approximately doubles learning compared to traditional lectures and laboratories. A research meta-analysis on undergraduate small-group science, math, engineering and technology learning found an average effect size shift from the 50th percentile to the 70th on content and large effects on retention in attitude. Other studies have indicated that merely increasing student in-class involvement with high-level cognitive responses to questions is correlated with critical thinking skills. Instructors often want to know what kinds of activities increase discussion. The answer is that involving students in virtually any type of activity generally increases discussion.

What else can instructors do to facilitate student-student discussion? Have students prepare for discussion by bringing a completed short assignment to class that will be the focus of a discussion. During the class discussion, raise questions that include student misconceptions. Place students in small discussion groups that are structured to promote discussion among all the students in each group. There are many good books and articles on how to optimally structure small groups for better in-class learning.[16]

When instructors first hear about constructivist and other active teaching techniques, one of the main criticisms is that these methods appear to cut down on "coverage" of material. After all, taking time for students to have discussion activities during lecture time means far less time for the instructor to lecture. Such concerns are legitimate; however, if instructors feel they need to maintain their level of content, then making effective use of student-instructor contact time could preserve the content. Rather then using lecture time for highlighting what students should know in the textbook, simply supply the students with practice exams that will cause them to read the textbook for explanations of the concepts. In addition, key questions can also be given such as "Summarize the nature of X" followed by "How would X differ if Y?" where Y is a commonly-held student misconception. For example, "Summarize the nature of mammalian evolution." "How would the nature of mammalian evolution differ if dinosaurs did not become extinct?"

Although moving some content coverage outside of class time is one way to deal with the coverage issue, instructors may want to rethink coverage altogether. Research results indicate that decreasing the volume of content increases understanding and long-term retention. Studies ranging from biology nonmajors to medical students show the benefits of a reduction in information density. The results of a comparison study of a low content biology nonmajors course to a relatively higher content biology majors course suggests that student understanding of the basic biological concepts was probably the same.[17]

Choosing textbooks is an important responsibility for the instructor. The National Research Council Report *Bio 2010* states that "Textbooks that bridge different disciplines and provide a coherent framework for study and learning can play a vital role in achieving the objective of more interdisciplinary and relevant biology education. Many high-quality biology textbooks are available."[18] It is up to the instructor or committee of instructors to spend time comparing biology textbooks for their particular courses.

Naturally there is more to increasing interdisciplinary teaching than the important step of choosing the proper textbook. Funding agencies and college administration will have to provide funding for new techniques, modules, and materials, and promote cross-departmental collaboration. The following may be useful to instructors who need federal-level recommendations to help encourage administration.

> *"Successful interdisciplinary teaching will require new materials and approaches. College and university administrators, as well as funding agencies, should support mathematics and science faculty in the development or adaptation of techniques that improve interdisciplinary education for biologists. These techniques would include courses, modules (on biological problems suitable for study in mathematics and physical science courses and vice versa), and other teaching materials. These endeavors are time-consuming and difficult and will require serious financial support. In addition, for truly interdisciplinary education to be achieved, administrative and financial barriers to cross-departmental collaboration between faculty must be eliminated."[19] National Research Council, 2003*

The Biological Sciences Curriculum Study (BSCS), a nonprofit corporation that develops and supports the implementation of innovative science education curriculum for students in kindergarten through college, has a learning model based on constructivism that can be incorporated into a great variety of curricula called the *BSCS 5E Model*. This method works best in smaller classes in which the instructor or teaching assistants are in closer contact with students, rather than in large lecture hall settings. The focus of the model is on the action of the students and is a way to engage students in learning. It was developed in the 1980s, has been used extensively, and has become a hallmark of BSCS programs.

The 5 Es are Engage, Explore, Explain, Elaborate, and Evaluate. As the BSCS explains: "First, students are **Engaged** in the concepts through a short activity or relevant discussion. Next, students **Explore** the concepts with others to develop a common set of experiences. In the **Explain**, the teacher guides the students as they develop an explanation for the concepts they have been exploring. In the **Elaborate**, the students extend their understanding or apply what they have learned in a new setting. In the **Evaluate**, the students and the teacher have an opportunity to evaluate the students' understanding of the concepts."[20] More descriptive information on how to implement the 5E model is available from BSCS.

INQUIRY INSTRUCTION

Inquiry instruction has been a very popular teaching method in recent decades. Like most methods there are variations on the inquiry process but they generally all share the characteristic of active learning in which students are instructed in a way that encourages discovery. This does not mean that instructors expect students to make all the great scientific discoveries on their own in a short period of time in a teaching laboratory—quite the contrary. Students are expected, though, to discover simple phenomena by means of activities developed for this purpose. It is pedagogically better for students to experience a phenomenon that to be told about it by an instructor.

Reportedly, some instructors have successfully used a form of inquiry learning in large lectures. To explain what they do, they use the analogy of a murder mystery. The instructor discusses the pieces of evidence, the circumstances, the timing, and so forth; and the students conduct an investigation that leads to their conclusions. Advocates of this type learning suggest that this approach is better (assuming that it leads to appropriate conclusions) than having told students the conclusions in the first place. Students had to put the pieces together themselves, which makes for a longer-lasting, deeper, and more engaging learning. The difference lies primarily in conducting the investigation oneself, rather than being told initially who "did it" and then being told the details.

In *guided inquiry,* instructors give students direction in their projects and investigations via focused questions, suggestions, and ideas. Sometimes the instructor acts as a supervisor for each team of students, and the students must report to the instructor before proceeding. The following is an example of a guided inquiry activity about leaf anatomy:

Inquiry step: Begin teaching a concept by showing students some natural phenomenon.

Sample activity: Have students observe and sketch a cross-section of a privet leaf.

Inquiry step: Ask students to observe the phenomenon and then speculate what caused it to happen or suggest its significance.

Sample activity: Students draw representative tissue samples and notice gaps between spongy parenchyma cells. Ask: are gaps formed in slide preparation (are they tears in leaf tissue) or are the gaps real and filled with water or gas? Could there be a vacuum between the cells?

Inquiry step: Choose one question and develop a hypothesis, prediction, or speculation based on the observations.

Sample activity: Students predict that the gaps are real and filled with liquid or gas.

Inquiry step: Design an investigation to test the hypothesis (or prediction or speculation).

Sample activity: Squeeze the leaf or heat the leaf and observe what comes out.

Inquiry step: Allow students to conduct the investigation to test the hypothesis, or provide them with information about such a test.

Sample activity: If students can't decide on a procedure to test their prediction, suggest that an intact privet leaf be placed in hot water—when this is done, tiny bubbles form, but only on the bottom of the leaf.

Inquiry step: Analyze and discuss the data. Allow students to draw conclusions about the hypothesis or prediction and the nature of the phenomenon.

Sample activity: students discuss ideas and suggest that the gaps in the leaf cross-section are real and filled with gas because bubbles came from the heated leaf.

Inquiry step: Apply newly gained knowledge to new situations.

Sample activity: The observations also suggest that there are holes, located mainly in the leaf bottom, through which gases can pass, but not on the top of the leaf. This leads to the discussion on diffusion and photosynthesis and should be linked to discussion of leaf anatomy.[21]

In *open-ended inquiry,* students have more independence in designing, conducting, and modifying investigations. They evaluate their results, come to their own conclusions, and report their findings. Instructors are involved in this process but to a much lesser extent than in guided inquiry. Open-ended inquiry is used when instructors think that most of their students can come to scientifically appropriate conclusions concerning a particular concept while working on their own.

In *collaborative inquiry,* students work on real, not contrived, investigations in which neither the instructor nor the students know the outcome. Environmental studies often present good opportunities for collaborative inquiry methods.[22]

LABORATORY

Traditionally, laboratory classes are not inquiry-based. Instead they follow a "cookbook" methodology in which the students follow specific directions in a lockstep manner with no discovery components. Cookbook labs have been shown to be relatively ineffective. Students are merely verifying something they have been told will happen if they "do" the lab correctly.

Instead of providing cookbook experiences, science lab classes should provide experiences with realistic scientific questions when possible. These laboratory activities should be structured so that students work as individuals for part of the time and in cooperative groups for most of the time. The laboratory experiences should increase in their difficulty as they progress, with increasing instructor/lab manual/TA independence where practical. One goal is to stimulate student interest and participation. Not only do project-based lab experiences help in this regard, but they can also be used for developing and improving scientific writing, speaking, and presentation abilities of students.[23]

The National Research Council (NRC) recommends that biology labs have interdisciplinary approaches to solving problems that include chemistry, physics, and engineering; the Council also recommends that chemistry, physics, and engineering labs include biology. The NRC proposes a three-step model for laboratories called the "crawl, walk, run" approach. During the crawl phase, students are provided step-by-step instruction and they record data. During the walk phase, students are provided guidelines and examples of how to carry out an experiment, and during the run phase students are given open-ended questions. The examples below of interdisciplinary laboratory foci using the crawl, walk, run approach have a biomedical emphasis and incorporate biology-related laboratory activities into physical science labs.

Physics laboratory

Crawl and walk phase foci

- *Conservations of energy*: energy input and storage, basal metabolism, measurement of energy expenditure, external and/or internal mechanical work, and energy efficiency.
- *Newtonian mechanics*: muscles as force actuators, moments created by muscles, free body diagram analysis within the context of human joint mechanics, ground reaction forces, mechanics of gait-running and standing balance, calculation of the center of pressure and center of reaction, inverse dynamics modeling of a simplified foot to determine ankle reaction forces, moments, and powers; and force control within the context of motor control.[24]

Run phase foci

- *Biomedical measurement*: cell, nerve, and muscle potentials; electrocardiograms (ECG), electromyograms (EMG), body temperature, control of body temperature, heat loss from the body, blood pressure measurement, body flow and volume measurements, noninvasive blood-gas sensors, optical microscopy, cell adhesion, optical sources and sensors, lung volume, heart sounds, drug delivery devices, surgical instruments, or electroshock protection.
- *Medical imaging*: origin of x-rays, the x-ray beam, attenuation and absorption of x-rays, x-ray filters, beam size, radiographic image, production of x-rays, computed tomography, ultrasound, MRI, nuclear imaging, single-photon emission computed tomography, or positron emission tomography.[25]

Engineering-for-life-scientists laboratory

Crawl phase foci

- Conservation of energy: energy input and storage, measurement of energy consumption, external and/or internal mechanical work, energy efficiency.
- Muscles as force actuators; moments created by muscles; calculating the point of gravity; force analysis of a system with one, two, and more joints;

set-up a mass-spring system. Attach an accelerometer to the mass and measure the response.

- Building a simple robotic system that can move and carrying out a mechanical analysis of the construct.[26]

Walk phase foci

- Building RC circuits; mimicking an action potential of a nerve cell; simple coupled RC circuits. Circuit analysis.
- Osmotic pressure versus hydrostatic pressure; building a cell; pressure measurements; analysis of systems with varying pore sizes and/or sizes of charged particles. Modeling the kinetics of charge separation.
- Visualizing and analyzing the path of charged molecules/particles in microfluidic devices. Experimentation and modeling.[27]

Run phase foci

- Building a human eye from optical components. Analysis of its performance; corrective optics for the human eye.
- Light sources and optical components (filters, lenses, lambda/And lambda/quarter plates, polarizers).
- Introduction to optical microscopy; illumination, building of a simple telescope or microscope.
- Differential interference microscopy.
- Confocal microscopy.
- Photophysics of light absorption and emission, competing deactivation pathways; kinetic analysis.[28]

Chemistry laboratory

- Synthetic organic chemistry experiment where different groups of students could perform the reaction at different temperatures to determine they rate constant for the reaction, and also its energy of activation, and for different times, to see the effect on yield of the product.
- Determine the effect of reaction conditions, such as the duration of synthesis, and the ratio of the desired product to other products.
- Introduce students to statistical analysis of experimental data when variation in results among students performing the same experiment occurs.
- Examine "unknowns" via infrared and nuclear magnetic resonance spectroscopies and students deduce their chemical structures, perhaps also being given a mass spectrum.

- The class can follow an enzymatic reaction using optical spectroscopy of quenched samples (so they do not tie up the spectrometers) at different times, but with varying pH's and/or the addition of inhibitors with varying substrate concentrations. This would let them determine and try to understand the rate laws involved in the reason for a pH dependence.
- Nonduplication of traditional established experiments, more project-based and real-life-like research goals to increase excitement.[29]

MISCONCEPTIONS

The National Research Council's Committee on Learning Research and Educational Practice has highlighted as a "key finding:"

> *Students come to the classroom with preconceptions about how the world works. If their initial understanding is not engaged, they may fail to grasp the new concepts and information that are taught, or they may learn them for purposes of the test but revert to their preconceptions outside the classroom."*[30]

The National Research Council's Committee on Undergraduate Science Education reports that "recent research on students' conceptual misunderstandings of natural phenomena indicates that new concepts cannot be learned if alternative models that explain a phenomenon already exist in the learner's mind."[31] These alternate models are, of course, misconceptions. This means that students must actively examine their initial conceptions and explicitly compare them to appropriate conceptions, otherwise their misconceptions are likely to remain unchanged and may even distort any related new material that is being learned.

Thus, instructors must first uncover students' misconceptions to more effectively teach. How might this be done? The following is an example from a nonmajors introductory-level general science class having over 100 students. The instructor shows students a battery, bulb, and a piece of wire (insulation stripped off at each end) and has examples to pass around in the auditorium. The examples that are passed around are inoperable (battery dead, bulb burned out) so students will not stumble across a proper way of lighting the bulb at this point. The instructor then asks the students to draw a schematic showing how to use the materials to light the bulb. While the students are drawing, the instructor quickly moves throughout the auditorium glancing at schematics being drawn and, if necessary, asks students to clarify their meaning. Within 10 minutes the instructor has some ideas of the students' conceptions regarding DC circuits. These ideas, plus those gleaned from prior reading on misconceptions common for students concerning simple DC circuits, and those ideas from the instructor's prior experience in conducting this exercise with similar students, now allows the instructor to follow the four basic constructivist-based teaching steps in the "essential elements of instruction" listed previously in this chapter. The instructor can now ask students to follow their schematics using working materials to make the bulb light (Step 1). If the bulb lights, then students may understand the concept of a simple DC circuit or they were "lucky." Knowing student misconceptions, the instructor can ask pertinent questions to determine which is the case. If the bulb does not light, the instructor can query students in a way that forces them to reflect on their circuit and propose alternatives (Steps 2 & 3). Students should be asked to light the bulb using their newly-developed alternative (Step 4). This procedure should be followed until the students determine how to light the bulb and show understanding of the concept of a DC circuit by applying their new working procedure to a new situation (e.g. two bulbs).

The contrast between understanding student misconceptions and conducting instruction in this constructivist manner, and teaching by telling without an understanding of student misconceptions is dramatic and very important. Not only does following constructivist-based teaching methods help directly confront students' misconceptions, but it also helps the instructor better understand at what level to instruct. One of the most difficult things for instructors is teaching at the right level for students' current understanding.

Many instructors assume incorrectly that if they uncover students' misconceptions, and if they properly use constructivist methodology during class time, then the misconceptions will be extinguished and will be replaced with scientifically accurate conceptions. However, extinguishing misconceptions is often very difficult and time-consuming. Most instructors report that properly addressing misconceptions almost always, for most students, takes multiple constructivist-type sessions over a period of months or longer to have the desired long-term effect. Many misconceptions may not be directly biological but rather, for example, may be physical or chemical misconceptions that underpin fundamental or complex biological concepts. In addition, there are misconceptions that have originated from non-formal education that often underpin students' trouble and learning biology. Students have had many years of making "sense" of their experiences and even the best constructivist-based teaching methods typically do not work right away.

Some believe that students will become irritated at instructors for attempting to determine their misconceptions. To the contrary, some evaluation research has shown that students perceive good courses as those where the instructor and staff make a real effort to understand their difficulties.

There are numerous types of misconceptions, mistaken assumptions on which those misconceptions are based, and origins from where misconceptions are formed. Five types of general misconceptions with examples related to evolution are shown in Box 2. Box 3 contains results of a study designed to develop a description of student conceptions about evolution. The subjects were nonmajor introductory biology students, most in their junior or senior year of college. The results revealed three major ways in which nonmajors' conceptions about evolution differed from the "population thinking"[32] based conceptions widely held by scientists. Similar results were found in a study of under-prepared college students enrolled in a biology course.

As the examples in Box 2 and Box 3 show, sometimes students' misconceptions are not so completely illogical when one understands how the student is thinking. A researcher who studied 322 university sophomores' understanding of natural selection concludes: "There seems to be a structure of misunderstandings which allows one to logically trace the origin of the misunderstandings to mistaken assumptions."[35] So detecting a biological misconception may not be the only problem because quite often the particular biological misconception is underpinned by more fundamental biological misconceptions and/or physical science misunderstandings.

CONCEPT MAPS

Concept mapping is a visual representation of the hierarchical organization of concepts. This technique has been used for decades as a way for students to relate concepts and learn how to organize content. In addition, instructors can use student concept mapping as a type of diagnostic tool to see how students perceive the structure of the content presented to them. While concept mapping is used in all subjects, much of the research involved in concept mapping involves biology lectures and labs.

How are concept maps developed? Key concepts are first selected and are then arranged in various ways, for example, from general to specific. Typically, a concept map links scientific concepts to form propositions about the concepts on the map. The following are some example propositions with scientific concepts first and last connected by linking words: mutations are a source of variation, environmental factors can influence survival, and vertebrate embryos have notochords.

Designs of concept maps vary, but generally concepts are written in circles or ovals. The concepts can be processes (e.g., photosynthesis), procedures (e.g., staining cells from a cheek swab), or even products (e.g.,

From-experience misconceptions are those that students surmise (consciously or unconsciously) from their everyday experiences. One from-experience misconception is that mutations are always detrimental to fitness.

Self-constructed misconceptions occur when information that students see or hear conflicts with what they already "know" (their misconception) but, rather than change their misconception, they accommodate the new knowledge in the framework of an old misconception. For example, students who have a long-held impression that evolution is predictably progressive, with the end goal being humans, will incorporate natural selection into that type of determinism. A deterministic perspective does not appear to be relegated exclusively to undergraduates or to nonbiology graduate students. In a study of 15 biology students at the master's degree level, the researchers noted a trend for students to propose deterministic-type explanations of evolutionary processes.

Taught-and-learned misconceptions are unscientific "facts" that have been taught informally by parents and others (and sometimes formally by school teachers) or unconsciously learned from fiction. For example, previous to college, numerous undergraduates have been told the Lamarckian idea of inheritance of acquired characteristics. Also, many repeatedly have seen dinosaurs and humans coexisting in print and visual materials such as films, books, and cartoons.

Vernacular misconceptions arise from the difference between the scientific use of a word and its everyday use, and the consequent misunderstanding of the distinction. More explanation and examples are given shortly, but one of the most prevalent examples of a vernacular misconception is evolution's status of being "only" a theory.

Religious and myth-based misconceptions are concepts in religious and mythical teachings that, when transferred into science education, become factually inaccurate. Two such misconceptions are that organisms do not have common descent and that the Earth is too young for evolution (and most geological processes) to have occurred.[33]

Box 2 Types of Misconceptions

carbohydrates). The superordinate concept is usually at the top with subordinate concepts flowing downward with increasing specificity, so the maps are generally read from top to bottom. Solid lines, dotted lines, and arrows connect the circles or ovals to illustrate linkage between one concept and another. The arrows are used to illustrate a non-bidirectional proposition.

1) *Students thought that the environment itself (rather than genetic mutation, sexual recombination of genes, and natural selection) causes traits to change over time.* They failed to distinguish between the separate processes responsible for the appearance of traits in a population (which originate by random changes in genetic material, e.g., mutation and sexual recombination) and the survival of such traits in the population over time (natural selection). Students had difficulty understanding the key idea that the environment affects the survival of traits *after* their appearance in the population. Instead, students thought that change was due to environmental forces alone, which acted on organisms to produce change, and held various conceptions about how the environment exerts its influence.

Some held teleological/Lamarckian conceptions; that is, students attributed evolutionary change to need. For example, students thought that if cheetahs needed to run fast for food, nature would allow them to develop faster running skills. Some held other Lamarckian views, attributing evolutionary change to use and disuse. For example, students thought that if cave salamanders did not use their eyes for many generations, their eyes would become nonfunctional. Also, a number of students attributed evolutionary change to organisms' ability to change in response to environmental "demands." For example, students thought that polar bears adapted to their environment by slowly changing their coats to white.

2) *Students did not view genetic variation as important to evolution, even though such variation is essential to evolution taking place.* Students viewed evolution as a process that acts on the species as a whole, not recognizing that it is variation among individuals within a population that constitutes evolution's raw material. They did not understand that the process of natural selection is dependent on differences in genetic traits and in reproductive success among individual members of a population.

3) *Students viewed evolutionary change as gradual and progressive changes in traits, rather than as a changing proportion of individuals with discrete traits.* Students did not recognize that traits become gradually established in a population as the proportion of individuals possessing those traits grows with each succeeding generation. They believed that all individuals change slowly over time. For example, students thought that if salamanders living in caves did not need sight, then they would pass down genes conferring less and less ability to see until the salamander populations were blind.[34]

Box 3 Some Nonscience Majors' Misconceptions about Evolution

Examples are designated by dotted circles or ovals. Linking words are usually used to help explain the connecting lines and form the propositions. Figure 2.1 is an example concept map about the design of a pencil.

Concept maps have been used extensively in the biological sciences. One example of the use of concept maps in a course on evolution is discussed in a study in which students were able to develop concept maps of class lectures, one per week for ten weeks, for extra credit.[36] The instructors briefly explained how to design concept maps and each lecture would provide approximately five seed concepts that were required to be on the maps. Students could then add five additional concepts and/or examples and hand it in the following week. The students who chose to do the mapping were in essence forced to study the material weekly instead of waiting just prior to an examination. Research results concerning the use of concept maps in this way indicated a positive correlation between concept map quality and course academic success. Although the results may simply be attributable to students who do better on exams doing better on concept mapping, there are other benefits from the maps. Students who made concept maps reported spending an average of 37% more study time on this college biology course than on their previous biology courses. In addition, constructing the maps forces students to explicitly structure their thinking each week, and by looking at the maps, the instructor can see where to adjust teaching approaches or content. In small classes or lab sections, it is

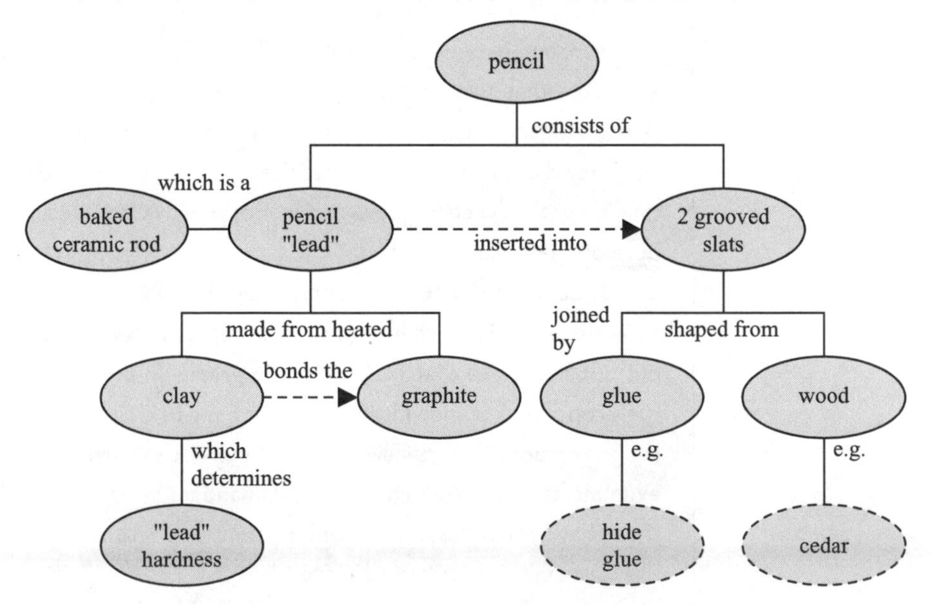

Figure 2.1 Example concept map about the design of a pencil.
Source: Wandersee, J. H. (1990). Concept mapping and the cartography of cognition. *Journal of Research in Science Teaching. 27:* 923–936. p. 933

helpful for students to compare maps with other students so they can see other possible connections. Research results suggest that developing concept maps help students remember key ideas but that they are not as good as taking and studying notes in helping students remember detail.

Another use for concept maps is to distribute instructor-made maps to students prior to lecture or labs. These maps can be used as advance organizers for lectures or labs, helping students see the general connections, thereby allowing them to concentrate more on detail. These organizers are intentionally structured at a high level of abstraction, meant to link and integrate the material while optimally increasing curiosity.

Concept maps are also used for exam purposes. On an exam students are to arrange given concepts in hierarchical order, showing relationships. Sometimes students are encouraged to add other concepts and examples as well. They are often graded on correctness of the hierarchy, propositions, examples, and linkage. For an exam on evolution, for example, students could be given terms such as environment, evolution, variation, speciation, adaptations, overproduction, fitness, and natural selection.

A strategy that helps students categorize and characterize information, often useful in disciplines like biology, is the development of *matrices*.[37] Matrices are like tables, and simply involve placing categories and their characteristics in a grid system. Categories (topics) can be placed along the top of the grid, with characteristics along the side, or vice versa. Information fills the cells appropriate for each intersection of a category and its characteristic.

Figure 2.2 shows an example matrix that categorizes information regarding protection, support, and movement in animals. In this case, the categories are animal phyla, and they form the left side of the grid. The characteristics are skeleton, musculature, and movement, and they form the top of the grid. The cells are filled with information pertinent to each of the three characteristics for each phyla.

After students understand the mechanism of matrices, there are a variety of pedagogical applications. The instructor may distribute pre-made matrices with the characteristics and categories in place. After lecture or lab, students can fill in the cells as a summarizing exercise. Once students become more adept with the matrices, the instructor may leave out either the characteristics or categories, with information in the cells in place. Some instructors eventually leave out two of the three components, leaving the students to fill in the other two. As with the concept maps, small group work during lecture and lab with matrices has been shown to be beneficial to both students and instructor.

Representative animal (phylum)	Skeleton	Musculature	Movement
Sponge (Porifera)	Gelatinlike middle layer supported by hard, sharply pointed skeletal pieces	No true musculature	Sessile
Jellyfish (Cnidaria)	Absent, but water in bell acts as hydroskeleton	Circular muscles in swimming bell	Propulsion: circular muscles in bell contract, pushing water out and moving the jellyfish forward
Flatworm (Platyhelminthes)	Hydroskeleton	Circular and longitudinal muscles underneath the epidermis	Waves of contraction that generate thrusting and pulling forces
Roundworm (Nematoda)	Hydroskeleton	Longitudinal muscles underneath the epidermis and cuticle	Movement by alternate contraction of muscles on either side of body
Earthworm (Annelida)	Hydroskeleton	Circular and longitudinal muscles underneath the epidermis	Alternate contraction of circular and longitudinal muscles results in peristaltic movement
Clam (Mollusca)	Shell (exoskeleton)	Muscular foot	Muscles in foot allow clam to burrow into sand or mud when shell is open
Insect (Arthropoda)	Exoskeleton (integument is skeleton)	Primarily striated muscle that attaches to exoskeleton	Ground movement by means of limbs; flight by movement of wings
Sea star (Echinodermata)	Hydroskeleton and endoskeleton of hard calcium carbonate plates	Retractor muscles in tube feet	Protraction and retraction of tube feet connected to the water vascular system
Human (Chordata)	Bony endoskeleton	Striated muscles attached to skeleton for locomotion; smooth muscle in organs; cardiac muscle in heart	Movement by lever action of limbs

Figure 2.2 The Skeleton, Musculature, and Movement of Representative Animals

LECTURING

Imagine an auditorium with over 100 students waiting for class to begin. In the front of the room sits a 8' high by 4' wide "box" with a black curtain covering it. There is string attached to the curtain going through a pulley at the ceiling with the string coming back down near a podium. The students think this is nothing unusual because the instructor typically has things covered until an appropriate point during the lecture. This time their teaching assistant walks in the classroom and states that "Professor X is detained so I will begin class. We will continue with our topic of buoyancy. Let's imagine that an adult human is underwater with approximately 5 lbs. of lead weight attached around the waist. What would bring the human to the surface? a) A plastic bag such as this filled with air [the teaching assistant holds up a plastic sandwich bag], b) A quart-size plastic bag such as this one filled with air [displays this bag also], or c) Neither bag is sufficient to bring the human to the surface." The options are also projected onto a screen.

The teaching assistant gives the students time to discuss the question and then opens the receivers for answers, while the students respond with personal response systems (transmitters). Approximately a third of the class chooses each response from the tabulations projected onto the screen. The teaching assistant then says "Well, let's find out." He pulls the string which in turn removes the curtain covering a large tank filled with water. The instructor of the course has been sitting at the bottom the entire time with his usual coat and tie (all soaking wet) but with a scuba air hose and regulator for breathing (air tank on the outside with a long hose). After the students' shock and laughter settle down, the instructor pulls a sandwich size plastic bag out of his sport coat and begins to fill it with air to try the options. The sandwich size bag does not work, but the quart size bag does work. The instructor floats to the surface and conducts the rest of the lecture standing in approximately 4 ft. of water dripping wet holding a waterproofed remote control for the PowerPoint presentation.

Could the students learn just as well without the theatrics? Are such lectures motivating or inspiring to students? Are these academic lectures or edutainment? What is a lecture?

Donald A. Bligh in *What's the Use of Lectures?* defines lectures as "more or less continuous expositions by a speaker who wants the audience to learn something."[38]

What are the elements of a good lecture? Gordon E. Uno, in *Handbook on Teaching Undergraduate Science Courses: A Survival Training Manual*, suggests that lectures should, among other things:

1. provide a model of scientific activity and problem-solving approaches—not just listing the results (facts) of scientific investigations;
2. explain, clarify, illustrate, and organize difficult concepts;
3. reveal to students methods of learning and thinking about difficult topics;
4. help students analyze information:
5. show relationships among seemingly dissimilar ideas;
6. formulate and help solve problems, and develop hypotheses and predictions;
7. evaluate and criticize evidence and alternative solutions;
8. relay current research and theories;
9. summarize material from a variety of sources;
10. provide structures to help students read and study more effectively;
11. help students integrate and retrieve information;
12. challenge beliefs, misconceptions, and habits of thinking;
13. impart new information;
14. promote interest and enthusiasm in the subject;
15. provide additional examples and illustrations;
16. help students answer major questions in biology;
17. provide opportunities for students to experience concepts in ways other than the laboratory and their textbook.[39]

Lecturing is the most common form of formal teaching throughout the world. The United States Department of Education recently reported that 86% of faculty members say that lecturing is their primary method of instruction. Nonetheless, traditional 50-minute sessions of *uninterrupted* lecturing is not the most successful way to facilitate learning. Interjecting brief episodes of active learning with structured student-student interaction, preferably constructivist based, will appreciably increase understanding and retention. For example, having students work in small groups to discuss and work out problems within the lecture period with the instructor visiting some of these groups throughout the classroom/auditorium is generally a significant improvement to facilitate learning. However, the reality is that many or most instructors—for a variety of reasons—choose to conduct traditional-type lectures without significant change to the exposition format of lectures. Therefore, this section about lectures will assume traditional lectures as the norm and will discuss how to improve traditional lectures without changing the format dramatically.

STUDENT EVALUATIONS OF LECTURES

Research over decades indicates that students' opinions are reliable, in the sense of being consistent, concerning rating a lecture. For example, when students listen to a lecture more than once by an instructor, the rating is very consistent. Likewise, when different students rate the same instructor give a lecture, the ratings change little. This is not an argument for or against students' opinions being valid but rather for their reliability of opinion. Answering the validity question is much more difficult, which is discussed in the student evaluation section of the next chapter.

If students' opinions had no significant correlation with increased understanding from a lecture, then many would question the validity of student opinions. Content surveys/exams given after a lecture "treatment" *do* show a correlation between students' rating of the lecture and their post-lecture content exam scores. A significant positive correlation exists between student-reported satisfaction for a lecture and content knowledge/understanding gain. However, if the "treatment" lecture is at a difficult level for half the students in the class as evidenced by the post-lecture exam, then these students may report feeling frustration due to struggling. Moreover, if the lecture treatment was at a relatively easy level, resulting in the lower half making an average-to-good grade, the upper-grade portion of the class may actually report dissatisfaction because they feel it was "too easy" or "I already knew that." All the average-to-good grade students report satisfaction because they finally "understand something in this class" or "I learned something in that lecture." This could cause a significant inverse relationship with grade and lecture satisfaction.

In addition to issues with student evaluation of lectures, there is the problem of varying distributions of *multiple intelligences* in courses. More than two decades ago, Harvard's Howard Gardner outlined a theory of human intellectual competencies that has since become a very important and immensely popular educational idea. It essentially posits that human intelligence is more than a single property, with "an intelligence" being defined as "a biopsychological potential to process information that can be activated in a cultural setting to solve problems or create products that are of value in a culture."[40] The nine identified intelligences are linguistic, logical-mathematical, musical, bodily-kinesthetic, spatial, interpersonal, intrapersonal, naturalist, and existential. Gardner's basic idea is that students learn better when their stronger or more predominant intelligence or intelligences are roughly aligned with how instructors engage students. For example, students who have strong narrational intelligences would likely enjoy, be intrigued, and do well with instruction that introduced evolution with illustrative stories about Darwin's voyages. However, students having relatively weak narrational intelligences, but having relatively strong quantitative/numerical intelligences, may not find such introductory instruction as enjoyable, intriguing, or profitable as introductory instruction that included population genetics problems.

Varying distributions of multiple intelligences in a lecture could create differing assessments. For example, if 80% of the course's students were those who learned well through stories (e.g., lessons that used the story of Darwin's voyage or the life course of a particular species) and 20% were students who were intrigued by numbers (e.g., population genetics problems), than a lecture that involved primarily quantitative/numerical aspects would most likely produce more favorable evaluations and higher post-lecture content exam scores than a lecture that was primarily narrational.

These examples illustrate that answering the validity question of a lecture has its complications due to, among other possibilities, variables among students. Numerous studies using class averages and large numbers in an attempt to even out individual variables have come to differing conclusions, such as (1) student ratings are a measure of teaching effectiveness, and (2) student ratings are not indicators of teaching effectiveness but rather of student satisfaction. One leading researcher's review of decades of studies concluded that "the chief correlates of students opinions of lectures are features of their own personalities, courses, and learning strategies."[41]

Since students' opinions appear to be related to the nature of the students, it would be by far more productive to change the lectures to better fit the students, rather than trying to change the students to fit the lectures. The latter might be doable, but it would take much longer than the duration of a course. In this regard, student evaluations about lectures, not just entire courses, are valuable. Naturally the evaluations of lectures from a previous course is helpful, but more helpful are evaluations given early and often in the course so that major and minor adjustments can be made with the same population of students. This not only helps the population of students who are being helpful in filling out questionnaires, but it also ensures consistency of the population when modifying the lecture "treatment." In addition, focus groups are extremely beneficial although, for many, they prohibitively time-consuming.

IMPROVING LECTURES

As discussed previously, traditional lectures are far less effective than most instructors think for a variety of reasons. However, lectures will remain a part of higher education for some time, because they are perceived—in comparison to other non-lecture teaching methods—as being (among other things) cost effective; time efficient with respect to "coverage," preparation time, and student contact time; empowering for the lecturer; and comfortable for the student because they need only to sit passively and take notes. If traditional-style lectures are going to be given for whatever reasons, it is best to do them well.

The traditional aspect is interesting because there is little evidence that today's lectures are any better than those of a century ago. Granted, teaching technologies have advanced greatly but has that improved the teaching techniques of lectures? When Charles Darwin started at Edinburgh University, with his brother Erasmus doing his external hospital study at Edinburgh, they were voracious readers, checking out the most library books of anyone that first term at the University. In addition, they purchased numerous books. Even though a "bookworm," Charles found nearly all lectures to be "dull" except for those given by a chemistry professor whom he considered to be entertaining—"chemical drama." Charles was not the only one; there were over 500 students in the course due to the professor's teaching style. Many instructors today think large numbers of students are a modern phenomenon. It was the largest class at the university and Charles considered it taught "with great *éclat*." Whether more than a century ago or today, few instructors lecture well.

There are instructors who criticize other instructors' classes as being too entertaining, contending that lectures should be serious places. Likewise, there are instructors of courses that students find entertaining, who criticize other instructors' courses as being dull and dry, contending that lecture should be educational and entertaining. Then, naturally, there are the majority of instructors somewhere in between these two positions. Wherever instructors stand on this continuum all would hopefully agree that lectures should not be unnecessarily dull, nor should they be entertaining without substance and rigor. So what should be the goal? Ideally, students should find lectures inspirational and stimulating. Some research has even suggested that motivation results in good study methods and may be at least as important as intelligence with regard to influencing students' achievements and learning.

What drives student motivation is much more diverse than just what increases excitement in the subject. Even though students have a great variety of motivational profiles, there appear to be some motivations that are quite common:

- wanting to enhance employability via qualifications;
- wanting to enhance self-efficacy through acquiring knowledge and skills;

- wanting to succeed within a competitive assessment regime;
- wanting to master a discipline/body of knowledge;
- wanting to please or impress others (parents, teachers and so on);
- wanting to learn for its own sake;
- wanting to play a role (student, engineer and so on);
- wanting to belong to a group (the university, the department, a coterie).[42]

Since most student learning occurs outside the lecture time, it is important to inspire and stimulate students so that they find the subject sufficiently interesting, therefore increasing learning outside of lecture. Some feel they inspire and stimulate by setting high standards in the course so that students strive for the points. This, of course, is a major motivator for numerous students and is extremely common and easy approach for instructors to implement. Other than the motivation in the pursuit for grades, how can one inspire and stimulate in a traditional lecture?

First and foremost is to do no harm—making things worse than the first moment the instructor walked into the room. For example, instructors should not frustrate students with illegible writing, projection of charts and graphs to small for everyone to read, sound problems (e.g., can't hear clearly in the back of the room), inability to operate audio-visual equipment, and so forth. These are areas in which students are clearly authorities to which instructors should pay close attention.

So what makes differences in lectures to students? Two general factors of clarity of subject matter and rapport consistently appear in published research. Here are common subfactors within the two general lecture factors:

1. Clarity of Subject Matter (more to do with subject matter than feelings)
   (a) organization
   (b) speed
   (c) clarity of its purpose
   (d) visual aids
   (e) summaries
   (f) effective use of time
   (g) lecture's knowledge of the subject
   (h) care in preparation

2. Rapport (more to do with feelings than subject matter)
   (a) personal style
   (b) questioning techniques
   (c) opportunities for discussion
   (d) inspiring interest
   (e) responsiveness to the audience
   (f) general personal characteristics or social skills[43]

Interestingly a few researchers feel that inspiring interest in the subject should not be a major objective of lecturing. They agree that there are some outstanding lecturers who do inspire students and their influence is long-lasting; however, they feel the overwhelming majority of instructors, if they inspire all, it is only for a very short time. The time is so short that the inspiration effect does not last more than a month or so. This being their case, they make the argument that student motivation should not be a major objective of a lecture. They certainly agree that it should be an objective of a lecture, just not a major objective such as transmitting information.[44] However, we contend that it is possible to increase instructors' inspiring factor and that student motivation should be a primary objective of the lecture.

Student arousal is important, although many instructors seem to incorporate few factors that affect students' attention. Variations in stimulation in the lecture situation are important in relation to the general level of brain activity to stay on task. Some of the more obvious, but generally not practiced ways to vary stimuli include varying auditory stimulation, visual stimulation (as in illustrations, demonstrations, etc.), posture (having to do with students' position, such as having them stand briefly, move, observe a demonstration close-up if in a small enough class), novel stimulation (every 15 to 20 minutes do something other than lecture as mentioned previously), and intensity of stimulation (not necessarily louder, sometimes much softer).[45]

Most instructors have heard about students' attention drifting 5–30 minutes into a lecture. Instructors who vary stimuli keep student attention longer than with no variation in stimuli, but holding students' attention for 30 minutes is very difficult. Pauses or short breaks in lectures have been shown to increase waning attention. The number of pauses or short breaks should be dependent upon how bored the class is at the moment. Lectures involving very difficult or very easy material, by the students' perception, tend to more quickly bore the students. These interruptions of the lecture can be small group interactions about the lecture material, individuals writing down lecture reactions, and so forth. Some researchers have reported that the break need have little-to-nothing to do with the subject material of the lecture to give positive results. In fact, they suggest that a brief rest from lecture allows students to return to lecture at the end of the break or pause to be at a much higher level of performance. The learning lost due to rest periods is small compared to the learning gained.

Recent reports on undergraduates show that a significant portion is sleep-deprived. Nevertheless, some studies have reported that students do better on immediate post-lecture tests in the morning than in the afternoon. Likewise, students do better on lecture/test experiments during midmorning than in the late morning. People are also more alert on Mondays and Tuesdays than on other weekdays. Naturally, none of this can be changed by a lecturer, but such information can be used to schedule more difficult-to-understand material at the most beneficial times.

To be motivating in a lecture an instructor must first understand what is motivating to the majority of students in the course. This is dependent on the level and kind of biology course, and what motivates students should be discovered by the instructors through interactions with students. There are, however, some rather common motives among students. Students want to know that the biology they are learning in a particular lecture has relevance not only to important biological concepts but also to their career paths. In addition, most students are naturally curious about the living world. The trick is to not to lessen or extinguish that curiosity with lecture.

In some cases, instructors can increase student motivation by increasing curiosity. In addition, many students are motivated by the fear of performing poorly and thus becoming non-science majors. However, some research results suggest that as student interest increases (due to non-fear related factors), the amount of fear-related factors necessary to motivate decreases. Does the "fear factor" significantly increase learning? Research is inconclusive, and, obviously, lectures should not be designed to intentionally frighten students. Moreover, some students are motivated more by positive encouragement than perceived threats. Praise and student self-esteem are important matters to keep in mind during lectures as they can make significant differences.

Many young students are motivated by social interactions with other students during breaks/pauses for small group discussions during the lecture time. Social interaction is not necessarily a direct motivation for these students; rather, having short social interactions during the lecture satisfies that need, allowing the lecture itself to become more motivating. Nonetheless, some reviews of the research report, with some reservations, that lectures are as effective as most other methods to "transmit" information, less effective than good discussion methods to promote thought, and ineffective to teach behavioral skills.[46]

Note taking has changed dramatically over the past decade due to various PowerPoint-type notes uploaded for students or distributed as handouts. The majority of studies indicate that note taking appears to aid memory and improve performance. Much of this can be attributed to keeping the mind "on task" when taking notes. Providing students with notes may result in many becoming overconfident about the lecture because they think that virtually everything they need to know from lecture will be on those notes. Conversely, students who feel they need to take notes generally have increased attention, resulting in greater concentration on the lecture. Possibly the best of both worlds is for the instructor to supply basic keyword and diagram/chart/figure notes prior to the lecture and tell students that these notes constitute less than 50% of what you need to know from this lecture. That way the basic structure and more complex artwork is there for the students as an outline of the lecture, but they need to pay close attention to taking additional notes to fill in the rest on their own.

SMALLER VERSUS LARGER LECTURE CLASSES

For biology classes, the evidence does not provide a conclusive answer. Certainly, larger classes are more challenging for the professor in many respects, such as attempting to get student feedback, giving individual student attention, maintaining students' attention, becoming familiar with a variety of students' personalities, and so forth. There are certainly ways to increase success in accomplishing these tasks in any class, but it is easier to do the smaller the class. Nevertheless, if class sizes must remain large, improving these factors will help improve student motivation and learning.

One of the positive aspects of classes with hundreds of students is that the instructor is usually assisted by teaching assistants. Therefore, the students have access to more than one instructor, and topics can be explained by multiple people in various ways. Not only does this give students the opportunity to learn from people having differing personalities, ages, academic backgrounds, and so forth, but it also often reduces the student stress level. Students generally feel that they can be more open with teaching assistants who are closer to their own age and academic level, than with the instructor of the course. Teaching assistants and students often become friendly in a way that is difficult for students and an instructor teaching a large course. Likewise, teaching assistants often can relate better to students having been in their academic place typically only a few years earlier.

Another positive aspect of large classes is that equipment is often available to support lectures that may not be affordable or practical for small class use. If an instructor wants to purchase a piece of equipment for a demonstration to 300 students, for example, the instructor has a much better chance of getting it funded than if the course has 30 students. Funding aside, equipment such as personal response systems are generally more useful with large classes rather than with small classes, because small classes may have too few students to obtain representative samples of a larger population, depending on the question. Also, error due to, among other possibilities, one student accidentally entering a response not chosen by the rest of a class of 30 would naturally have a much larger representation than the same one-person error in a class of hundreds. Clearly there are many of these types of advantages to large lectures.

But small classes do have advantages, of course. Although some students are extroverts, having little-to-no problem voicing their ideas to be challenged in the largest of auditoriums, most young undergraduate students are reticent to speak in a large class. In small classes, with proper encouragement, support, humor, and care, most students will take more risks during discussions than they would ever take in large classes. This situation is very positive for learning and can never be equaled in auditoriums.

There are a great variety of learning and teaching strategies in addition to lecture, for example, discussions, debates, individual/team research, AV, demonstrations, explorations (e.g., dissections), data analysis, idea analysis, computer interactions/educational technology, cooperative group learning, problem solving, inquiry activities, lab activities, field activities, and more. Instructors who want to do much more than previously discussed may want to have students instruct portions of lab or discussion sections themselves. It appears many of us learn extremely well when we must teach, and so do many students. The common relevance to other methods is illustrated in a popular education generalization about how most people seem to learn:

We learn:
10% of what we read
20% of what we hear
30% of what we see
50% of what we see and hear
60% of what we write
70% of what is discussed
80% of what we experience, and
95% of what we teach.

ADVICE FOR NEW INSTRUCTORS

Veteran instructors often know precisely what topics they will explore and how much time is needed for each in their courses. Therefore they can have a list of topics on the course outline that very accurately reflects what topics will actually be taught during the semester. New instructors rarely have the experience necessary to calculate this prior to teaching a course a few times. Merely asking a senior instructor who has taught the course many times about topic coverage is not foolproof because that instructor may spend different amounts of time on the topics than a new instructor does. If you are a new instructor, it may be useful for you to state in your course outline "these topics are tentative; additional topics may be added while others on this list may not be covered." Even some senior instructors often place the tentative topics notice on their outlines because they like to see how that particular semester or section of students understands the various topics before proceeding to others, thus adapting the topics midcourse. Dates of exams, quizzes, papers, lab reports, and so forth should have exact dates on the course outline because many students feel the need to plan, or at least understand, the sequence of evaluations in the course. Having "TBA" for an exam on a course outline, or worse, changing a date for an exam on a course outline, is extremely frustrating to numerous students and is often against regulations.

New instructors often find themselves most nervous when they lecture, especially to large classes. This should not be thought of as unusual because instructors report that lecturing causes them more anxiety than any other part of the professorial job. Over a few months or years the stress will diminish, but newer instructors should use their stress to help demonstrate a higher level of energy in class. Students can often detect a lack of excitement in senior instructors that they see in newer instructors. Instructors should appear if they are enjoying themselves. The students will see the apparent enjoyment and the energy, with only the instructor knowing that some of the energy may be nervous energy. So nervousness may not be a negative attribute after all.

In addition to demonstrating excitement about the subject and the class, instructors can reduce the stress of lecturing by reducing the amount of time they lecture. One way to do this is to punctuate the lecture with small-group discussions. Not only is this technique pedagogically more effective than lecturing continuously for 50 minutes or more (as discussed previously), but it allows the instructor some breathing space. This "timeout" from lecturing can be used to review notes, get back on track if that section of the lecture was not going well, or visit members of the class. Walking around the auditorium to the very back row, visiting and talking with members of the "audience," not only can relieve significant levels of stress for instructors but also can increase the students' attention who sit in those distant back rows. Rather than being annoyed, students often feel an increased sense of belonging to the class and a closer connection with the instructor.

Whether stressful or not, one does not like forgetting what is next in the lecture or putting something out of order. Large numbers of instructors use PowerPoint-type presentations for the obvious reasons, and for the less obvious reason of a note guide to remind them what to do or talk about next. If the presentation is well constructed, instructors need no other form of notes or reminders than the presentation itself.

Let your university help you hone your teaching. Many universities and colleges now have official offices or centers for improving teaching and learning. Many have a secondary goal of conducting research in higher education teaching and learning, not only to make improvements at their institution, but also to add to the research base. In either case, these units generally have people who have expertise in teaching and learning, and who can likely answer questions not only about pedagogy, but also about equipment-related queries such as availability of personal response systems (PRS), their use in relation to class-size, and so forth.

CHAPTER

3 Assessment

"...and after this comes judgment [or assessment]."
Hebrews 9:27

There is a great variety of assessment methods, approaches, tools, and philosophies. Nevertheless, with whatever methods an instructor uses to assess student achievement, students should have good reason to think that the instructor has a sufficient amount of data to accurately judge their level of understanding at the completion of a course. Students should believe that they were assessed fairly, and that the assessments gave them a chance to indicate their true understanding. Fair assessment can mean a few things: an instructor accurately marking assessments throughout the course, students agreeing that the exams, quizzes, and labs were fairly marked, and instructor's assessments being valid and reliable measures to get at students' true understanding.

GOOD EXAMS

A good exam supplies evidence of the level of student understanding on the topic for which the assessment was intended. Such an exam is not easy to construct. Most instructors use test banks supplied by publishers, author their own questions, inherit exams from others, or piece together exams from various sources that include their own questions. Typically, instructors attempt to refine exams over time by rewording, removing questions, adding choices, and so forth. But are these "good" exams?

What if an instructor slightly revised his or her exam questions and a student who previously received an A now received a B, and a student who previously received a B now received an A? What if exams or quizzes used during the course were re-administered to these two students six months later without giving them time to study, and they both got the same grade? What if the students were allowed to take the exam on a computer? Would their grades rise significantly because they can write and think better when using a computer than when using paper and pencil? Would some students do significantly better if they were given more time? If such variables significantly alter students' grades, then are we testing biological understanding or biological understanding when constrained by time and environmental factors? If the latter, then are the exams "good?"

Measurement error creeps in to all sorts of assessments, not just biology exams, of course. When data is collected on an exam it does not necessarily reflect the true understanding of the student that the instructor would like to measure. Instead, the data (i.e., the grade) reflects not only students' biological understanding, but also a long list of variables that affect the students' performance on the exam. Some variables that affect student performance, in addition to the factors already mentioned, include sleep deprivation, emotional stress, and time of day. Measurement quality can also be affected by instrument flaws, administration flaws, random fluctuations in subjects over time, rater or scorer disagreement, and other matters.

If students' understanding has truly changed, however, some research contends that they should then act differently than they did before their understanding changed in contexts related to that understanding. Assessments become more about how students perform in unfamiliar or novel contexts and less about picking or writing a correct answer. If students have not performed differently but have picked or written correct answers on assessments, then chances are very good that the same students given the same assessment months later will do dramatically worse. However, if students *have* performed differently, then chances are very good that the same students given the same assessment months later will do well. Assessment tasks requiring student performances of understanding at non-declarative levels are constructivist in nature and can better confirm what students really understand.

Rather than having students merely repeat what they know in examinations, many instructors would like to see students apply their knowledge to novel situations. This is also an excellent way to focus on higher-order thinking skills using biological knowledge. The following are some selected questions from an introductory botany course for nonmajors. An important point is that the instructor did not directly discuss any of the following questions in lecture or lab. The students were expected to apply their knowledge to new situations.

- What kind of plant parasite do you think might kill its host sooner: a green parasite or an orange one? Choose one, and explain your answer.
- Do you think you have any genes that are identical to that of a plant? Explain your answer.
- If you had a tall plant and a short plant, how could you tell if the tall or short was the dominant characteristic?
- A student from last semester designed the following experiment. One bean seed is planted in each of three identical pots, with the same soil, and placed in the same location. No salt is added to Pot A, 1/2 teaspoon of salt is added to Pot B, and 1 teaspoon of salt is added to Pot C each week. Explain what is wrong with this experiment's design.
- On a quiz in lecture you saw a mouse and a plant in a glass container — and both remained alive! Explain how the situation is similar to the fish in the big, sealed-up, glass jar in the lab.
- Suppose you bring home a Christmas tree and put it in a bucket of water. The water level in the bucket never changes, and the leaves of your tree start to fall off the tree. What do you think is happening and why?
- 100 lettuce seeds were placed on moist filter paper in a glass dish in a warm (25°C) sunlit room. Another 100 lettuce seeds were placed on moist filter paper in a glass dish in a warm (25°C) dark room. Seeds in both dishes germinated. Based on how this experiment was designed, which do you think is more important for germination: the moisture on the paper, the glass dish, or can you tell? Explain.
- A student added nitrogen to some corn plants and no nitrogen to others. Only one corn plant with nitrogen lived, but it grew 2 feet tall. Nine corn plants without nitrogen lived, and the averaged 1 foot tall. Based on these results, can we conclude that nitrogen helps plants grow tall? Explain.[1]

Another approach to assessment in higher education is to use methods that better mirror research processes if the students have been exposed to research instruction. The ways in which students will study is affected by the ways in which they will be assessed. Thus, the methods used to assess students should be aligned with the goals of greater research awareness, understanding, and abilities. Some sample suggestions from Jenkins, Breen and Lindsay in *Reshaping Teaching in Higher Education: Linking Teaching with Research* include:

- "Make the task one that has students simulate writing an article for a learned journal (and using the effective assessment criteria for that journal), or writing a consultancy report.
- Make the task one where students (perhaps in groups) make a bid proposal to carry out a research project. Here the assessed task is likely to emphasize the research design and project plan.
- Ensure the assessment criteria support developing student understanding of research and knowledge of research methods.
- As academics we generally show our work to others, and then revise it before submitting it to a journal or research granting authority. So students can be required to have their work reviewed by others in the class. These student reviewers are then trained in how to referee and review research articles, with particular attention to research based assessment criteria such as the appropriateness of the research methodology. The final version might include a commentary by the student on how he or she had taken into account (or decided to reject) the views of his or her peers.
- Have students present their work at a simulated research conference (possibly with external observers and assessors, for instance, researchers from another institution or professional body).
- Have students present their work for assessment as contributions to a journal or edited collection on a topic. Web-based publishing allows this to be done both cheaply and publicly: and enables students (and staff) to get feedback on their research from researchers worldwide."[2]

ASSESSING BIOLOGICAL LITERACY

The Biological Sciences Curriculum Study (BSCS) in *Developing Bological Literacy*, suggest characteristics of students at four levels of biological literacy. The same student may not be at the same level for different concepts; the student may be one level for natural selection and another level for cellular respiration. These levels do not imply that all students must reach the highest level for all concepts; some lower levels are appropriate for some concepts at this point in their education. This framework may be helpful for instructors and may also be helpful for students to better anticipate what sort of biological literacy characteristics exist, and what levels the instructor may expect.

Nominal Biological Literacy
Students can identify terms and questions as biological in nature, but possess misconceptions, and provide naive explanations of biological concepts.

Functional Biological Literacy
Students use biological vocabulary, defined terms correctly, but memorize responses.

Structural Biological Literacy
Students understand the conceptual scheme of biology, possess procedural knowledge and skills, and can explain biological concepts on their own words.

Multidimensional Biological Literacy
Students understand the place of biology among other disciplines, know the history and nature of biology, can ask and answer their own discipline-related questions, and can understand the interactions between biology and society.[3]

ACCURACY OF GRADING/MARKING

Can we claim that one method of instruction is better than another because students do better on exams after experiencing that method of instruction? One can imagine the classroom as a laboratory in which to conduct an experiment with the subjects being the students, and the treatment being the instruction. How can you measure if or how much the students learned from the instruction? What relationship exists between the instruction and students?

One of the problems in carrying out such measurement, and one reason that it is not an easy nor a straightforward process, is threats to validity. The following table shows some of the complications involved by listing four broad categorizations of threats to validity: statistical conclusion validity, internal validity, construct validity of cause and effects, and external validity.

Some instructors, often those in the physical sciences, feel that quantifying can eliminate virtually all problems in higher education research. However, the almost absolute confidence once held in the traditional quantitatively measurement-based reliability and validity has been significantly shaken. Many contend the assumptions underpinning the traditional measurement model are questionable or faulty. Filling some of the confidence void in assessment has been the increasing agreement among educational researchers about reliability and validity in qualitative assessment. In today's educational research it is fair to state that instructors should place value on instructors' professional judgments of quality, while being at least suspicious of so-called objective measurement-based instruments that are reliable and valid. There are many books and articles on measurement errors that may be helpful. For combining the strengths of both qualitative and quantitative methods into multiple-choice science tests see Sadler.[4]

Table 3.1 Threats to the Validity of Designs

Threats	Features
(1) Statistical conclusion validity	Was the study sensitive enough to detect whether the variables covary?
(a) Low statistical power	Type II error increases when alpha is set low and sample is small; also refers to some statistical tests.
(b) Violated statistical	All assumptions must be known and tested when assumptions necessary.
(c) Error rate	Increases, unless adjustments are made with the number of mean differences possible to test on multiple dependent variables.
(d) Reliability of measures	Low reliability indicates high standard errors which can be a problem with various inferential statistics.
(e) Reliability of treatment	Treatments need to be implemented in the same way from implementation person to person, site to site, and across time.
(f) Random irrelevancies in	Environmental effects which may cause or interact with setting treatment effects.
(g) Random heterogeneity of	Certain characteristics in subjects may be correlated with respondents dependent variables.
(2) Internal validity	Was the study sensitive enough to detect a casual relationship?
(a) History	Event external to treatment which may affect dependent variable.
(b) Maturation	Biological and psychological changes in subjects which will affect their responses.
(c) Testing	Effects of pretest may alter responses on posttest regardless of treatment.
(d) Instrumentation	Changes in instrumentation, raters, or observers (calibration difficulties).
(e) Statistical regression	Extreme scores tend to move to middle on posttesting regardless of treatment.
(f) Selection	Differences in subjects prior to treatment.
(g) Mortality	Differential loss of subjects during study.
(h) Interaction of selection with	Some other characteristic of subjects is mistaken for maturation, history and treatment effect on posttesting; differential effects in testing selection factors.
(i) Ambiguity about direction	In studies conducted at one point in time, problems of of causality inferring the direction of causality.
(j) Diffusion/imitation of	Treatment group members share the conditions of their treatments treatment with each other or attempt to copy treatment.
(k) Compensatory equalization treatments	It is decided that everyone in experimental or comparison of group receive the treatment that provides desirable goods and services.
(l) Demoralization of	Members of group not receiving treatment perceive they respondents are inferior and give up.

(continues)

Table 3.1 Threats to the Validity of Designs (continued)

Threats	Features
(3) Construct validity of cause and effects	Which theoretical or latent variables are actually being studied?
(a) Inadequate explanation of Constructs	Poor definition of constructs.
(b) Mono-operation bias	Measurement of single dependent variable.
(c) Mono-method bias	Measurement of dependent variable in only one way.
(d) Hypothesis-guessing	Subjects try to guess researcher's hypothesis and act in a way that they think the researcher wants them to act.
(e) Evaluation apprehension	Faking well to make results look good.
(f) Experimenter expectancies	Experimenters may bias study by their expectations entering into and during study.
(g) Confounding constructs and levels of constructs	All levels of a construct are not fully implemented along a continuum, they may appear to be weak or non-existent.
(h) Interaction of different	Subjects are a part of other treatments rather than of an treatments intended one.
(i) Interaction of testing and Treatment	Testing may facilitate or inhibit treatment influences
(j) Restricted generalizability	The extent to which a construct can be generalized from one study to another.
(4) External validity	Can the cause and effect noted in the study be generalized across individuals, settings, and occasions?
(a) Interaction of selection and treatment	Ability to generalize the treatment to persons beyond the one studied.
(b) Interaction of setting and treatment	Ability to generalize the treatment to settings beyond the one studied.
(c) Interaction of history and treatment	Ability to generalize the treatment to other times (past and future) beyond the one studied.

Reprinted from Hisen, T. & Postlethwaite, T. H. (1994). *The international encyclopedia of education* (2nd ed.) from vol. 4 *Evolution of Evaluation Design* by Michael, W. B., & Benson, J. (pp. 2079–2089) with kind permission from Elsevier Science, Ltd. The Boulevard, Langford Lane, Kidlington OX5 IGB UK.

INTER-RATER RELIABILITY IN GRADING "SUBJECTIVE" ASSESSMENTS

Sometimes instructors evaluate students using assessment tools other than true or false, multiple-choice, and fill in the blank exams. For lack of a better term, such assessments will be referred to here as "subjective." If the class is large, teaching assistants often help grade subjective assessments, such as short answer questions or papers. When multiple raters for the same exam items are used, some degree of variance is natural when answers are dependent upon subjective judgment. Students know or at least suspect this, and often complain that multiple raters or scorers do not grade equally among students, and they question the meaning of course grades under these assessment circumstances.

One way to promote inter-rater reliability is to simply assign no more than one rater (teaching assistant) per exam question; however this is often not feasible in large classes. Therefore, the criteria that the instructor has created for scoring each question should be detailed sufficiently to reduce error so that it does not unnecessarily compound any already existing imprecision associated with the exam. In addition, have the raters discuss the questions and their answers, preferably with the instructor present, to agree on the distinctions between the letter or number grades, or whatever grading system is being used. Next, the TAs should discuss the distinctions between pluses and minuses between each letter or number grade. If the grading system includes finer distinctions, then the TAs should discuss these distinctions as well. Written criteria and discussions among raters will help standardize the subjective assessment process and increase inter-rater reliability.

To determine the extent of inter-rater reliability, the instructor should select a range of student answers to one question — one that the instructor thinks is excellent, one that is above-average, average, and so forth, until a rough range exists (e.g., one A, B, C, D, and F; or one 5, 4, 3, 2, 1). Make a copy of every level for each rater. The raters should grade the student answers without any communication with the other raters and should take approximately the same amount of time they will actually use when grading. Now inter-rater agreement and reliability can be better judged and fine-tuned by the instructor.

It is useful to explain to the class that the raters have graded the papers using standardized criteria specifically developed for each item. Letting students know about the precision involved should cut down on complaints about this matter and give students more confidence that the process is fair.

STUDENT-PERCEIVED QUALITY OF ASSESSMENT

An area of concern often brought up by students in course evaluations is the quality of their assessments. Usually students are not upset simply because they did not get the grade they wanted, but because they believe that assessments were inconsistent, incomplete, unfair, and inaccurate. If the instructor can better help students understand the assessment process and how it is consistent, complete, fair, and accurate, then the students will be more apt to see how and why they missed the mark, why they should take responsibility for missing the mark, and why their criticisms about the quality of assessment are not appropriate.

Instructors should find various ways to evaluate student understanding that both they and their students find appropriate. This does not mean letting the students control examinations, but it does mean having a variety of evaluation evidence that virtually all students would agree was sufficient to provide a clear picture of their understanding. If students still complain about the assessment process being unfair, then the instructor can ask students to suggest changes to evaluation procedures that would better reflect student understanding. The goal is to become an educational researcher who attempts to find out, albeit qualitatively, what might be changed to improve the quality of course assessment. Students may have some useful ideas.

CLEAR COURSE GOALS

Many colleges and universities now require instructors to write the goals of the course on outlines/syllabi. This is done primarily to clarify goals for the students so they better appreciate course direction. Although most students seem to understand what material will be "covered" on upcoming exams, they lose sight as to what key concepts they are to master. One technique is to point out which course content aligns with which course goals (e.g., "In better understanding this concept you'll be partially fulfilling goal #3 on our course outline.") Some instructors choose to write non-specific goals or a single course goal such as "To increase understanding of evolution." Certainly that is the goal of an evolution course, but this goal is too vague to give students much direction. It is important to provide students with some indicators (subgoals) that, when satisfied, would show that they have increased their understanding of evolution. If you do not articulate specific goals for students, then try asking them at the end of the course to state five or six (unwritten) subgoals of the course. If the answers from the overwhelming majority of students do not match your unwritten subgoals, then it is fairly safe to assume the students did not comprehend the goals. If you do articulate some specific goals for students and think that you have made these goals clear throughout the course, ask students the same question. If one goal is consistently absent or incorrectly stated, then it is evident where some work must be focused.

GOALS AND THE HIDDEN CURRICULUM

Out of the three general areas of biology, chemistry, and physics, biology students are known to be "the memorizers." Chemistry and physics are thought to be more conceptually and problem based while biology has a "reputation" of being primarily about factoids. This unfortunate but popular characterization seems to have faded somewhat over the past two decades, but much of it still remains. Some research results indicate that biological science undergraduates in their third and fourth years lack a satisfactory grasp of the fundamental biological concepts in spite of passing conventional assessments. They appear to remember and perform satisfactorily for course examinations yet appear not to understand the biology they are supposedly learning.

One common reply about the situation is that student "learning" is certainly occurring, but it is not so much about understanding biology as it learning how to adapt to instructors' requirements. Some time ago at MIT it was written that the institution's curriculum contained excellent educational goals including such things as critical thinking, analysis, originality, and other noble objectives. However, the students perceived a hidden curriculum that, in essence, indicated to students that they were to figure out how to best perform on the assessments, appease the instructors and teaching assistants, and look good in labs. They learned ways to achieve high grades, which were not necessarily ways to best learn. Two different curricula existed: one perceived by the instructors and the other perceived by the students.

The way to lessen this problem is to design assessments that better measure the extent to which the instructors' curricular goals have been met. It sounds obvious, but instructor assessments often do not test student attainment of course goals. One of the most common instances is a goal of understanding in which the assessment can be answered by merely memorizing material. Designing assessments so that mere memorizing will be of little help is the instructor's challenge. Using questions such as those listed in the previous example for a course in introductory botany are examples of these types of questions.

INTELLECTUAL CHALLENGE

Instructors' conceptions of what students are capable of achieving have been argued for decades in hallways, offices, and department meetings. Many senior instructors will say that students are not performing at the level they did years ago. Less senior instructors may say that they cannot give their students more challenging material because they simply can't handle it. New instructors in quiet moments of honesty may admit that less rigor in the course means spending less time with students and having more contented students, assuming that having more contented students translates into better course evaluations. Many contend that the quality of biology instruction in the whole department is being lowered at the undergraduate level.

Nonetheless, research shows that consistently high academic expectations are related to high levels of student performance. So if instructors follow what the research suggests, and keep or even raise the level of rigor of their courses (within reason), students will rise to the occasion, work harder, and academic performance should increase. However, this scenario presents many potential and interrelated problems.

First, instructors who increase the academic rigor of their courses must be willing to spend more time helping students, since demands for instructor help will likely increase. Second, instructors must be willing to allow students' grades to drop until students understand the rigor of the course and start working harder. This change in student behavior may not take place within a single semester, but it may take more time until the word gets around to students that these courses are difficult. Grading "on a curve" is not useful, because then it is difficult to effectively set a bar. Students can remain in the same distribution without rising overall, with the top grades given to students who might never have achieved beyond previous levels. Third, if grade inflation has occurred in other courses in the department, then it will be difficult to hold the students to a much more rigorous standard. Students will likely complain to administration that these courses are unreasonably difficult and therefore unreasonably time-consuming compared to other courses in the biology department.

With regard to the first concern, the increase in instructor time to help students is not as dramatic or proportional as one may think. Nevertheless, most instructors are overloaded with various duties and any further subscription of their time sounds unattractive. The first recourse is to set up more teaching assistant hours that are convenient for the majority of students in the course, such as immediately before or after lectures and laboratory sessions. Electronic help forums are always an option as well. In the final time tally however, instructors probably will have a modest increase in the time devoted to helping students, but this increase should be well worth the increase in student performance.

Instructors may want to speak with their administration concerning the advantages of increasing student performance. When administration agrees that this is certainly a noble goal, instructors may want to negotiate some release time from other duties for the increased instructor-student time, and increased funding for teaching assistants' extra time. Instructors should also discuss with administration and other instructors the potential problems of increasing academic rigor of one course while other courses may remain relatively less rigorous, if in fact they are. What might be the negative ramifications? Should the department consider increasing the rigor of multiple courses simultaneously? These are all difficult questions to answer and primarily have department-specific answers. In principle, the value of increasing the intellectual challenge in courses should outweigh administrative difficulties.

When raising the level of difficulty of a course, provide students with clear, appropriate goals and the structure to succeed in meeting them. Also, modify the course so that it will be more appealing and interesting to students. Students will be much more motivated to work hard in a difficult course if the course is interesting to them. Ask students what they find interesting in the course and brainstorm with them ways to make it more interesting without sacrificing rigor.

Instructors will probably learn a great deal when talking to students about their course concerns and suggestions, but the most informative talks will probably occur after the course is finished, grades are recorded, and students are now more open to discussion. For input on these issues during the course, enlist the aid of teaching assistants. Students generally perceive TAs as less threatening than the instructor, so they might more easily query the students and report back to the instructor. Ideally, this experiment may last over many years, but the ultimate result should be increased intellectual challenge with more student interest.

STUDENT INDEPENDENCE

Students learn in various ways. Instructors should not assume that they know the best way that students will learn. Individuals can come to the same level of understanding via different learning methods, and the method that is best for one student may not be best for another. The challenge for the instructor is to engender a sense of independence and control in students for choosing how to investigate the subject matter. In their course evaluations, students often express an appreciation for independence and choice in their courses. The instructor should promote student inquiry to the extent that over-dependence on instructors and teaching assistants is nonexistent. At least one instructor said he feels like a six-foot fluorescent yellow highlighter because students, with notes and textbook dutifully in hand, constantly ask him " is this going to be on the exam?" or "Is this important to know?" While it may give students a greater sense of security, such dependence is detrimental to student inquiry and interest in the subject.

GIVING FEEDBACK TO STUDENTS

Up to just a few years ago there was a leading research biologist at a North American university who would write the following on undergraduate student papers: "This is a waste of my time." "All wrong!" "You don't understand!" "You should have at least read the paper!" Not only were these types of comments on numerous pages of the students' work, the red pencil would scratch out entire paragraphs — one after the other — without a bit of helpful explanation as to why they were crossed out. The only hint was the scrolling in the margin stating "Wrong!" Such feedback to students is not only discouraging, it does not help the student improve. The students are not only depressed about their apparent substandard work but mystified about what to do. Instructors such as this are hopefully on the right tail of the severity curve for feedback.

Multiple reviews of research results in student evaluations reveal that poor feedback is one of the most commonly mentioned criticisms college students have of their instructors. Because feedback is so important to students, this is an area that deserves improvement — not only for them but also for increases in student evaluation ratings.

One obvious way of understanding the academic difficulties of students and how to communicate with struggling undergraduates is to spend time with them. Meeting with small groups of students is useful to discuss course material along with their difficulties, as is meeting individually with students who are particularly struggling with the class. If the class is large, then TAs could conduct discussion sessions with the instructor "floating" from session to session. The primary purpose of these meetings and discussions is for the instructor to learn what the students truly understand and how they communicate about it. Having 5-10 question-and-answer rounds face-to-face with students helps to better understand what they understand and what they don't. The instructor will often be able to tell by the look on their faces, along with the next question, whether the student is increasing in understanding or is perplexed. Over time, instructor communication should improve, resulting in better feedback to students, engendering better biological understanding, which in turn should cause students to feel they have had more effective teaching because of the feedback they receive.

COURSE EVALUATIONS AND ACCOUNTABILITY

Course evaluations are used primarily for improving courses and assessing instructors. It is debatable whether students have any expertise about what constitutes a good course, but evaluations are valid indications of students' thoughts (as long as they are being honest).

Are student opinions worth anything with regard to improving a course? The answer is "yes." Student opinions are valuable as an indicator of how motivating, interesting, and stimulating they find a course. While the instructor may view the course as motivating, interesting, and stimulating, the students may not. Instructors who do not know students' opinions or do not value them have little hope of improving their courses in this regard.

For the more subject related matters, student opinions are also valuable for instructors. After all, even if the majority of students are dead wrong about something in the course, such as believing the exams possess "trick questions," if the instructor does nothing to dissuade this false belief, the students will become dissatisfied and motivation will decline. To help correct the students on this matter, instructors must first know there is a matter that needs correcting. Instructors who are apathetic about student evaluations may find that they have to respond to administrators who have received student complaints. For many reasons, as surely most instructors will agree, it's better to teach students and respond to their concerns than to respond to administrators' concerns.

SUCCESS IN TEACHING?

Professors who have good course evaluations generally think that they are successful in their teaching. Those who have poor course evaluations generally think that the evaluations are not good indicators of teaching success but merely reflect instructor personality and grading practices. Some instructors think that if the instructor and the course are organized, tests are fair, homework is manageable, and the instructor appears enthusiastic, that the evaluations will be favorable. Many others think that they and their courses have these attributes, but they receive course evaluations that are lower than average.

Are student evaluations useful? Does responding to student concerns improve the course and subsequent evaluations? Paul Ramsden, in *Learning to Teach in Higher Education*, notes that students often prioritize characteristics of good instruction in the same way instructors do. For example, both students and instructors think that characteristics of good instruction include organization, understandable explanations, stimulation of student interest, clear goals, feedback on work, empathy with students' needs, and encouragement of independent thought. Neither think that instructor personality and sense of humor are as important.[5] In addition, research results also show that students can tell the difference between "attractive" teaching without substance and "non-attractive" teaching with substance. In other words, punching up bad teaching with a few bells and whistles does not significantly alter student evaluations of the course — contrary to popular belief.[6]

One aspect that often appears to engender positive student comments on evaluations concerning course instruction is the ability to redo homework, exams, tests, and so forth. Students speak highly of courses in which the instructor lessens the tension by allowing students to redo assessments by using different versions of the assessments. Whether stressed or not, students think this is a better learning environment in which they can see by the assessment, and their performance on it, where they need to improve. This practice reduces the grade distribution within a course, but the overall goal of the course is to improve student understanding. Thus, instructors may need to sacrifice grade distribution for greater student understanding. Using this approach to assessment will significantly increase correcting time, of course, and is probably one of the reasons why it is not commonplace.

Are you one of those instructors who think the student evaluation questionnaires you are forced to use are poorly worded or just do not ask the proper questions? There are alternatives. One alternative is to design your own evaluations and distribute them at various times in the course as a type of formative evaluation. Their purpose is to help you improve your instruction during that course and subsequent courses. You can then report the results as an empirical study to the Chair or Dean to illustrate the disconnect between the instructor-designed evaluation and the institutionalized evaluation.

One goal of student evaluations is to determine if the department has good instructors. However, the persons making this determination – the students – often vary in their expectations of instructors. For example, at some institutions education majors are sometimes very critical of biology instructors compared to biology majors. This may be attributed to education majors' training in what constitutes good pedagogy, so they have greater expectations. In addition, students rate some disciplines as having better teachers overall than other disciplines. An Australian study consisting of approximately 4500 students at 50 institutions looked at student perception of differences in teaching quality by field of study. The categories included good teaching, clear goals, appropriate workload, appropriate assessment, and emphasis on independence. The results indicated that the student perceptions of the relative quality of teaching varied by field — and the differences were quite large. The students rated the teaching in the fields of health sciences, engineering, law, math, and computer sciences as below average; in the natural sciences as about average; and in social science, education, humanities, and the visual arts as above average.[7]

Such results may make instructors feel as if it's not instructor variance but subject variance that really makes the difference. This may have some credence across school disciplines, but the Australian study also found that there were significant differences overall within subject areas and disciplines. For example, excellent, average, and poor examples of teaching could be found within every discipline.

Are students qualified to judge good teaching? Some educational researchers report that there is a fairly good correlation between factors that comprise good teaching and what students describe as a good teacher. These researchers also say that the students are uniquely qualified to judge instruction, because students experience various types of instruction, know what works for them, and are politically inert because their evaluations are anonymous. Nonetheless, instructors and administrators vary greatly within and among institutions on the value of course evaluations. But the goal remains: to increase instructor understanding of the relative effectiveness of their teaching/course as perceived by the students, and subsequently to make adjustments that will improve the course in both the eyes of the instructor and the students.

ACTING, INTERPERSONAL SKILLS, AND PERSONALITY

For many instructors, improving teaching is just a matter of following research-based principles about non-social matters. As a result, teaching improvements will likely occur and student achievement will likely rise. However, there is another piece to the teaching puzzle that many instructors find the most difficult to implement: the acting skills necessary to rise above negative feelings to make the instructional arena appear upbeat, inspirational, interesting, and so forth. The negative feelings may have to do with having had a bad day at home, or giving the same lecture 3 times in a row. It may be illness that creeps up during the course, or a student or two who are extremely annoying. Yet you must overcome these feelings and "act" the part of an interesting and engaging instructor.

Good teaching also requires patience, when sometimes you might prefer to treat irrational requests or complaints with brevity and a slight edge of condescension. Summarily dismissing students in this way may be acceptable in some students' views, and it is certainly expedient for the instructor, but whether right or wrong, many students today find such actions to be irresponsible, unprofessional, rude, and just plain mean. Students who are offended in this way might have increased problems in the course because their feelings about the instructor will interfere to some extent with their learning. On the other hand, some students might use those feelings to their advantage — to show the instructor that they can do well in spite of their condescension. Nevertheless, whether such instructor behavior increases or decreases student learning, instructors will "hear" about it in course evaluations. It's hard for students to feel inspired, motivated, and upbeat about a course when the instructor's personality or method of interacting with students is one they strongly dislike. Therefore, instructors should reflect on student evaluation comments about personality and social behavior and consider changing where prudent and possible.

The following characteristics are some that students identify with good instructors. This list is from the *Handbook on Teaching Undergraduate Science Courses: A Survival Training Manual.* Notice how many of these characteristics have something to do the instructor's social skills.

1. alert, appears enthusiastic
2. appears interested in students and activities
3. cheerful, optimistic
4. self-controlled, not easily upset
5. has a sense of humor
6. recognizes and admits own mistakes
7. is fair, impartial, objective, and patient
8. is knowledgeable
9. shows understanding in working with students and is sensitive to students' personal and educational problems
10. is friendly and courteous to students
11. commends effort and praises work well done
12. encourages students to do their best
13. organizes classroom procedures well, but is flexible with over-all plan
14. stimulates pupils through interesting and original materials and techniques
15. conducts practical demonstrations and gives clear explanations and directions
16. encourages students to work through their own problems and evaluate their accomplishments
17. disciplines in quiet, dignified, and positive manner
18. gives help willingly
19. foresees and attempts to resolve potential difficulties
20. is an effective questioner and listener, encouraging widespread response from students[8]

One of the most difficult issues where social skills are a must is the teaching of evolution in introductory majors and non-majors' courses. Many instructors believe they have little-to-no problem with students concerning evolution because they do not hear complaints. Anonymous surveys and focus groups with students after such courses often reveal that many students are deeply disturbed about evolution instruction. Students in introductory courses have an incredible array of backgrounds, which affect how they perceive the instructor, the instruction, and the subject itself. Their public high school backgrounds can range from having accurate and thorough evolution instruction to not having the word "evolution' ever mentioned in the classroom! Private high school students may have even been taught that evolution is not only false but is religiously undesirable and even dangerous.

No other biologically related subject, including stem cell research, abortion, euthanasia, environmental issues, and the like, can have such an effect on some students. After all, not only does it make students question where they came from, and then possibly what their life means, but to many it also questions their belief in an afterlife and their relation to a supreme being. Most biology instructors certainly do not attempt to engender such questions when teaching evolution, but a large number of students will begin to question silently anyway.

Naturally, it would be helpful for students to better understand that science does not consider supernaturalism, but introductory level students often believe that one type of accuracy cannot contradict another type of accuracy. So if those students perceive science and the supernatural as having conflicting answers, then the job of teaching science to those students becomes more difficult and necessitates good interpersonal skills. These students may lack motivation and interest in science because of their perceived religious conflicts with what they are being taught in the classroom. They may also have their concentration split between the scientific and philosophical/religious implications of evolution. Certainly there is a place for extra-science explorations to study some of these issues. Some biology instructors think that the introductory college biology classroom is an appropriate place, while others do not. Nevertheless, whether or not appropriate or practiced, many students will have conflicting feelings engendered by a science course. Some instructors think that they can teach evolution without offending any student. That is an impossible task.

Some students are offended just by the suggestion that they may have evolved from nonhumans. Generally, such students do not feel angry; they feel sad. They feel sad about what is being taught, sad that the other students may believe it, and sad for the instructor for thinking it is accurate. Others are not so sad, but still take offense. For example, one instructor who understands extremely well the sensitivities concerning students and evolution and has studied the sensitivities for decades, thought he had given a perfectly nonoffensive treatment of evolution in an undergraduate biology lecture. As he expected, he did not hear a single complaint from over 200 students in the course — until the anonymous course evaluations. Even though that lecture had been given months previously, students were still sensitive to the issue. The instructor was most surprised by the student who wrote "Concerning the evolution lectures, as a Christian I felt like you were mocking Christianity." This comment came from a highly capable third-year student in a highly competitive Research 1 level university. After reading this comment, the instructor listened carefully to the recorded evolution lectures, and he still could not determine what he did to cause the student's concern.

All instructors must remember that what we say and do is not isolated; it interacts with what the students bring to the course. In this case, it is likely that the student is a Christian who is a biblical literalist who finds offensive any assertion that humans are related to other organisms. The student probably believes that humans, and probably all organisms, were supernaturally created and did not evolve from common ancestors. Therefore, any teaching to the contrary may cause a perception of threat or mockery to this belief system. If a student "knows" X is absolutely true and an instructor teaches something that the student considers as anti-X, and X has to do with a student's perceived personal relationship with a supreme being, then there will most likely be feelings of discontent.

Instructors cannot know everything students bring with them to courses, and even if they did, there are many issues that cannot be avoided. After all, students learning evolution must learn relatedness of organisms. Nevertheless, what can be done is to hand out questionnaires at the beginning of the course about such eventually sensitive matters as evolution and have students fill them out anonymously. The questionnaires cannot require that the students write responses, because many students think that they give away their identities in their handwriting. The questionnaires may help give instructors insight into the sensitivities of their students, and they can then better prepare instruction. For example, there may have been ways to help the student just mentioned. The course could have been prefaced with how scientific questions and answers are different ways of knowing from non-scientific questions and answers. Even though science may have one answer and some people's perception of religion another, science cannot adjudicate those religious answers.

Often students feel threatened because they perceive science as giving the one and only answer that attempts to undermine their religious beliefs. Students who come to understand that it is not the goal of evolution instruction to change peoples' religious beliefs, often feel more comfortable with the instruction. Some students can come to understand why scientists conclude that evolution is the best scientific explanation for the diversity of life, while still maintaining that their religious beliefs hold otherwise. This may sound strange to many instructors but this situation helps many students feel less anxiety, cynicism, and stress engendered by the subject, the course, and the instructor.

Unfortunately, many instructors find it difficult to empathize with such students and feel the students should just "suck it up" and stick to the nuts and bolts of the course. Easier said than done; it would be like asking biology instructors to sit in a course where the instructor and textbook taught that biology was a soft, unworthy, inaccurate science compared to real sciences such as physics and chemistry. Imagine if that course even went so far as to claim that the evidence for descent with modification was nonexistent. Most instructors would probably have strong feelings about such assertions in an undergraduate course. This is roughly how many students feel about being taught evolution.

Instructors need to let individual students who have such concerns know that their instructor is aware of feelings engendered by the subject and is willing to talk with students during office hours. If instructors are willing to do this, it will go a long way for many students. Some students will have genuinely sincere questions about the evidence for evolution; others will have well-prepared arguments; others may bring in anti-evolution literature. Whichever the case, instructors will learn a considerable amount about students' misconceptions concerning the evidence for evolution. Moreover, instructors will also learn some of the interpersonal skills necessary to better help students with their various struggles in learning the science of evolution. (For more help in this area see the next chapter and the book *Defending Evolution in the Classroom*[9].)

CHAPTER

4 Evolution VS. Creationism

They cannot learn modern biology until they unlearn their intuitive biology, which thinks in terms of vital essences. And they cannot learn evolution until they unlearn their intuitive engineering, which attriutes design to the intentions of a designer.
Steven Pinker, Harvard University[1]

Most biology instructors have probably heard about the decades of national poll results typically indicating that approximately one in two Americans think the diversity of life is *not* due to evolutionary processes. Various polls show various numbers, but the data basically indicate that about 50% of Americans reject evolution. As an alternative to evolution, some Gallup polls have items about religious aspects such as indicating that God created humans pretty much in the form as they are today less than 10,000 years ago.[2] Other polls, conducted by scientific organizations such as a National Science Board, showed a mere 45% responding affirmatively to "countrywide quiz" items about humans having developed from earlier species of animals.[3] Other polls report that only about a third of respondents think the theory of evolution is supported by evidence.

Some polls ask about evolution understanding. In general, of those who recall ever having heard the term evolution, only half choose the correct layman's definition. Other polling items ask questions regarding the use of the term *theory* in science. They report that two-thirds of respondents agree that evolution is commonly referred to as "the theory of evolution" because it has not yet been proven scientifically. Only about one-third thought that evolution was completely or mostly accurate, while the remaining two-thirds responded that evolution was mostly or completely inaccurate, or they were not sure.[4]

Biology instructors find it hard to believe that millions of people reject evolution, think that dinosaurs and humans coexisted, and think that humans did not develop from earlier species of animals.[5] Instructors often tell themselves that persons who think this way must not be college educated. Yet national polls report that many respondents who hold bachelor's degrees also hold evolution misconceptions, and that the rate of rejection of evolution and holding evolution misconceptions is similar among college graduates and non-graduates.

A common comment of postsecondary biology instructors is that all the fuss on "our" side is basically a waste of time. After all, those instructors say, we are never going to change the minds of those who reject evolution anyway, so we should not try. People do not want others messing around with their afterlife, and maybe we shouldn't be doing that anyway. But why do virtually all people today affirm that the earth revolves around the sun and not vice-versa? At one time many people thought a heliocentric view was "messing around with their afterlife" because it made sense for them that God would have placed the earth as central because of its perceived importance. Over time, scientists and educators have been highly successful in increasing people's understanding of the accuracy of heliocentrism. Astronomers have done their educational job, and biology instructors can do the same regarding evolution.

Regardless of the success of astronomy educators, it is important to try to raise understanding of biological evolution as a scientifically accurate concept. Students complete chemistry courses understanding that the science is accurate, regardless of their grade in the course. They know the accuracy of balancing a chemical equation. Likewise, students complete physics courses appreciating that $F=ma$ is accurate no matter whether they received an "A" or "C" in the course. These are fundamental concepts in chemistry and physics. Yet large numbers of students, who had apparently taken biology courses with the full range of passing grades, find evolution—the fundamental concept and overarching theme of biology—to be scientifically inaccurate. They think—for a great variety of reasons—that the science of evolution is inferior to the rest of the biological sciences and is much less accurate than fundamental concepts in other sciences such as chemistry and physics. In a national poll about half of Americans asserted that evolution is "far from being proven," while, by comparison, the same poll reported that less then 10% think that Einstein's theory of relativity is "far from being proven."[6] While the word "proven" may not have been the best word to use in this polling, it nevertheless illustrates the difference in public opinion about these two sciences.

Yes, it is true that balancing chemical equations and understanding relationships between force, mass, and acceleration do not infringe on many people's perceptions of the hereafter. However, it is disconcerting that people feel that evolution is far less scientifically credible and has less evidence supporting it than other scientific concepts do. Many people think that evolution is "only" a theory, or a theory in crisis, and that scientists commonly question and debate whether evolution actually occurred.

Studies have shown that religious beliefs appear to interfere with the understanding or acceptance of scientific views, especially evolution.[7] Nevertheless, biology instructors should be teaching the facts of evolution and the reasons why evolution is considered factual, whether or not students' inaccurate ideas are engendered by religious beliefs. Teaching evolution is not an attempt to change students' religious beliefs; there should be no such goal in science education. At the end of a course in biology, however, students should realize why scientists conclude that evolution is scientifically accurate and understand the scientific processes and evidence of evolution, even if they reject evolution for strictly religious reasons. *Most of the work of biology instructors lies in correcting a host of misconceptions about evolution and its underpinning concepts.*

ORGANIZED RESISTANCE AGAINST EVOLUTION

Most biology instructors do not know that professional anti-evolutionists are well organized, well funded, and highly effective in their work. A majority of biology professors seem to characterize creationists as being nothing more than some backward-thinking people on par with flat-earth advocates, existing only in the Bible Belt, holding their beliefs solely because it says so in the Bible, and knowing nothing about evolution and related concepts. In addition, many biology professors think that creationists would not want to understand science even if given the chance, do not argue about anything scientifically-based other than to say that evolution is wrong and the Bible is right, and rarely go on to higher education. Those who do go on, they think, don't major in the natural sciences and definitely do not major in biology. These characterizations of creationists are *all* inaccurate.

The reality is that those who spend their professional lives promoting creationism are almost always college educated and often have degrees in the natural sciences—some with doctoral degrees in disciplines such as paleontology, biochemistry, and biology. They author what they perceive to be research articles in antievolution "scientific" journals. They write books about why the science of evolution is bad science and how logic dictates a creationist conclusion. They attend annual and present papers at international conferences on creationism. The only reason their work is not accepted into mainline peer-reviewed science journals, they contend, is because of anti-creationism bigotry among the evolutionary establishment and not because of poor science. They are quick to point out that there are numerous practicing scientists who are creationists, some of whom hold positions at respectable research universities. Naturally, the work they do at these universities is not explicitly related to attempts to discredit evolution. However, some creationists that hold appointments at private religious colleges and universities *do* work explicitly to discredit evolution and promote creationism.

In addition to rejecting evolution because of their religious convictions, most college and university-educated creationists contend that they reject evolution because the evidence for evolution is unconvincing and is simply bad science. Many go further and argue that the only reasonable explanation for life on earth is supernatural or extraterrestrial creation, and that this form of supernatural or extraterrestrial causation should be included, or at least considered, in the realm of science. They believe that the data simply lead to such conclusions, and that most or all scientists would come to the same conclusion if they were not blinded by naturalistic philosophy or religious teachings about evolution.

Creationists contend that their "good science" is better than the scientific community's "bad science." Creationists' "good science" goes by many names such as "abrupt appearance theory," and the oxymoronic "scientific creationism" and "creation science." There are many types of creationists, as we will discuss, and they hold a variety of differing views about science. Nevertheless, most creationists agree that evolutionary theory is seriously, if not fatally, flawed. Moreover, since much of the lay public think that life arose in one of two ways—by evolution or creation—most creationists present arguments against evolution hoping to illustrate that the scientific conclusion must be creation by default.

The specific arguments against evolution vary greatly by the type of creationism, but the following are some general examples of misconceptions typically held by creationist students.

1. Biological life could not have developed from the inanimate via natural processes.

2. The diversity of life we see today could not have evolved from lower forms of life.
3. No evolution can occur beyond, roughly, a phylogenetic level of "family."
4. Humans did not evolve from lower animals and, since their creation, have always possessed all the basic human characteristics of today's humans ("self-image, moral consciousness, abstract reasoning, language, will, religious nature, etc."[8]).
5. The earth is not old enough for evolution to have taken place.
6. Radiometric dating methods are inaccurate.
7. Most sedimentary rocks containing fossils are the results of a global flood occurring less than 10,000 years ago.
8. When originally supernaturally created, all organisms were created perfectly and over time have experienced physical degeneration.
9. The Second Law of Thermodynamics precludes evolution because entropy, not complexity, increases.
10. The gaps in the fossil record illustrate that evolution did not happen.
11. It is statistically impossible that life arose from nonlife by itself.
12. Evolution is *only* a theory; there is a good percentage of the scientific community who contend evolution is impossible.
13. Organisms look too well designed to have evolved.
14. A partial eye, wing, or other structure are of no use.
15. Dinosaurs and humans coexisted.[9]

In the past few decades, numerous creationists have developed formidable institutions for the sole purpose of combating the teaching of evolution at all levels. Two of the largest anti-evolution institutions in the United States employ over 100 full-time people, many of whom have degrees in the natural sciences. These institutions carry hundreds of anti-evolution book titles along with numerous videos, CDs, and pamphlets. Some creationist periodicals have a circulation of over 300,000 per month, and broadcast via radio network to over 1,500 outlets. They have extensive web sites reporting millions of visitors each year. The California-based Institute of Creation Research (ICR) has an accredited graduate school offering M.S. degrees in Biology, Geology, Astro/Geophysics, and Science Education. It also has a museum of "earth history" which reports an annual attendance of nearly 25,000 adults, children, and school groups. Answers in Genesis (AIG), the other large creationist organization, is currently raising funds to build a museum. Together, ICR and AIG have budgets of approximately $10 million annually.

In the last couple decades creationists have formally debated science professors hundreds of times at U.S. colleges and universities, often drawing large crowds.[10] There has been some evidence of a correlation between student attendance and occasional local anti-evolutionary meetings and challenges to the teaching of evolution.[11] Many biology instructors think that antievolutionism is solely American. This is not correct; creationism in its various forms is growing in popularity in many major portions of the world including Europe, Asia, in the South Pacific.[12]

More and more, biology instructors are becoming aware of the significant threat that well funded and well organized anti-evolution organizations have on the teaching of evolution at the secondary level. They realize that these organizations have an untold effect on students who will be involved in their colleges and universities in various majors outside of the natural sciences. They also realize that many potentially capable college students do not go into the natural sciences, particularly biology, because of what they have "learned" about the inaccurate, biased, anti-religious, and anti-God nature of evolutionary biology. They did not want to spend their time learning fairy-tales disguised as science.

Many postsecondary biology instructors were awakened to some of the potential effects of creationists on science education when the international news reported in 1999 that the Kansas State Board of Education successfully voted to remove almost all mention of evolution from the state's education standards and assessments for public schools. Instructors knowledgeable about the potential threats to science education at all levels have become involved in multiple ways to help answer the attacks of professional creationists.

Some college biology instructors understand creationist threats and are willing to become involved to promote evolution education, while the majority thinks that creationists are no threat to biology education. A third group of biology instructors have some understanding of the organization, funding, resolve, and possible effects of creationist organizations and subgroups on evolution education, yet these instructors are resigned to letting those who think that evolution is scientifically incorrect to "wallow in their ignorance."

[10] It is not recommended to formally debate professional creationists because it increases the publicity and credibility for antievolutionism. For example, one author (Brian Alters) has sparred with prominent creationists in front of large audiences and, consequently, the press has given the creationists coverage. For example, a newspaper reported one of these events as "Scholars Debate Creationist Theories at Harvard University" (Rechler, 1999). There, the author and renowned evolutionary biologist Graham Bell formally clashed with the world's leading creation evangelist, Ken Ham, and one of his scientific associates, Russel Humphries, physicist at the Sandia National Laboratory. Yet while we discommend providing a forum for creationists, if such a forum will be provided to them anyway (regardless of whether opponents participate), then science should be represented.

Some of these instructors have known and watched the anti-evolution forces grow in numbers, funding, buildings, museums, publications, debates, radio broadcasts, videos, web sites, personnel, and so forth for decades, but they have become apathetic because they think there is little-to-no way to change creationists' ideas about evolutionary science. Thus they contend that we simply should not try. Such a stance is counterproductive. While President of the AAAS, Stephen Jay Gould retorted concisely when he wrote in 2001:

> *"This battle must be won, but we cannot prevail (or at least not cannot prevail honorably) unless we meet our creationist questioners by grappling with the diversity of arguments, and with respect to the sincere and important reasons behind their misunderstanding of material that properly belongs within the domain of science, and cannot threaten the essence of religion."*[13]

TYPES OF CREATIONISM

Intelligent design

The newest and most "academic" type of creationism known is intelligent design theory (ID). Intelligent design theory is also sometimes referred to as initial complexity theory or theistic science. The new creationism of ID holds that evolution is an insufficient explanation for the diversity of life on earth, and that science should recognize that an intelligent designer or designers—whether supernatural or nonsupernatural extraterrestrial—created life on earth. Irreducibly complex biological structures must have been designed because evolution by natural processes cannot account for them. Instead, some unidentified form of supernatural or extraterrestrial intelligence designed complex biological structures such as DNA and the bacterial flagellum.

These ID views are not necessarily connected to any particular religion or belief in the supernatural. Common descent is often permissible by ID adherents although it varies by advocate. The primary difference between ID and other types of creationism is that ID advocates state that the identity of the designer(s) is (are) not a necessary part of the teaching of ID. However, it is probably not coincidental that virtually all prominent ID advocates are Christians.

Intelligent Design has been increasing in popularity over the past 15 years. It is often billed as "a scientific alternative to evolution." Other types of creationism have not gone away; rather it seems most student misconceptions based on creationism in secondary and introductory level college classes come from those associated with young-earth creationism (to be discussed). However, the new ID creationism appears to be an increasing problem in more advanced higher education partially because the discussions are often more advanced and philosophical in nature. ID is also an increasing problem in the governmental realms of school boards and state boards of education, because it attempts to distance itself from creationism, which courts have ruled as being religion and the scientific community has thoroughly discredited as being science. By distancing ID from a creationist label, ID advocates presume there is a better chance to get its tenets included in public secondary school science classrooms and fulfill an immediate goal to change instruction about the basics of biology. Also, ID advocates are able to create a large "tent" that represents almost all forms of creationism by being unspecific in many of their creationist claims.

Like the two young earth creationist institutions mentioned previously (ICR and AIG), ID advocates are well organized. The major ID group is headquartered in Seattle, Washington, with prominent members coming from diverse backgrounds, such as the law faculty at UC Berkeley and the biochemistry faculty at Lehigh University. Their budget reportedly exceeds $1 million annually. ID advocate activities include producing web sites, videotapes, and media publications; organizing conferences; authoring trade books; making media appearances; giving college and university talks; testifying at school board and legislative meetings; and giving talks at religious organizations. Some researchers consider ID to be creationism's Trojan horse because "design theorists" are using a "Wedge Strategy" which attempts to substitute "theistic science" for natural science in educational politics at all levels and regions.[14] Attempts have been made to influence state science standards and even influence federal education legislation. Recently, two outstanding researchers have painstakingly analyzed and documented the ID plan of action and report:

> *"Through relentlessly energetic programs of publication, conferences, and public appearances, all aimed at impressing lay audiences and political people, the Wedge is working its way into the American cultural mainstream. Editorials and opinion pieces and national journals, prime-time television interviews, and other high-profile public appearances, offhand but highly visible negative judgments on evolution or "Darwinism" from conservative politicians and sympathetic public intellectuals (assisted in their anti-science by a scattering of "feminist epistemologists," postmodernists, and Marxists)—all of these contribute to a rising receptiveness*

to ID claims by those who do not know, or who simply refuse to consider, the actual state of the relevant sciences."[15]

As with other forms of creationism, ID has no scientific data or testable theoretical ideas to contribute to the scientific enterprise thus far. Most science instructors think there is nothing to worry about. The danger, however, is that ID strategy is not actually aimed at the scientific community—it is aimed at nonscientists, such as elected officials. The nonscientific public hears about PhDs in the natural sciences, mathematics, philosophy, and other disciplines stating that evolutionary science is in trouble and that there is a new respectable scientific alternative called intelligent design theory. From coast-to-coast, the *New York Times* to the *Los Angeles Times* reports that the new creationism has "a more sophisticated idea" and more "academic respectability." This is what the public reads, and this public includes those who teach our students at the secondary level both inside and outside of formal courses. Those are the students who bring related misconceptions into higher education biology courses, which makes the job of postsecondary instruction more difficult.

Literalist

Literalists are the polar opposite of ID creationists. While ID advocates have a large tent that includes almost any kind of creationism, the biblical literalists believe only in the literal reading of the creation account in the book of Genesis. That being the case, they believe that the earth is young, somewhere between 4000 to 10,000 years old, that the days in the first chapter of Genesis are literal solar days, and that all life was supernaturally created essentially in its present form in the past 4000 to 10,000 years. They believe that the Bible is divinely inspired and that every portion is considered God's revelation to mankind. That is, the Bible does not just contain the word of God; it *is* the word of God. When it comes to science, just as with other matters, literalists believe the Bible is infallible, inerrant, and authoritative. Adam and Eve were the first two humans and all other organisms were created instantaneously within 24-hour days in the same week God created planets, stars, and light. Literalists are often referred to as young-earth creationists, and it is out of their ranks that most of the leading anti-evolutionary organizations arise.

Young earth creationists—even professional young earth creationists—accept some forms of evolution. They, however, would state that they do not. They are not being dishonest; they just define evolution differently from accepted scientific definitions of the word. Young earth creationists believe God created all organisms within certain "kinds" of living things as stated in Genesis ("each unto his own kind"). They believe that animals can change within those kinds but are limited by the parameters of the created kind. The creationist literature is not specific on this matter, but some leading creationists estimate these limits to be no greater than the phylogenetic level of family. For example, cumulative changes resulting in descendents that are roughly the same are perfectly acceptable, such as dogs into other types of dogs, which is an example young earth creationists would typically use. Therefore, literalist creationist leaders and higher education students have no problem understanding and appreciating the evolution of mutating pathogens, pests, and infectious diseases. They can understand why the scientific community concludes this science is accurate, but they cannot understand why the scientific community concludes macroevolution is accurate. There is no trouble with lessons concerning changes over time resulting in different species, genera, and sometimes families because they perceive these to be merely changes within a created kind. The problems begin when they perceive any large-scale evolutionary change to be change above the created kinds.

Progressive

Like the literalists, progressives believe that all life on earth was supernaturally created, but they believe that the universe, including earth, is much older (millions or billions of years) than what literalists contend. Progressives differ in essentially three ways concerning time and evolution. One faction of progressives holds that all life was supernaturally created essentially in its present forms in the literal solar days of Genesis within the last 10,000 years. Another faction disagrees and contends that the days in Genesis are indeterminately long, maybe even millions or billions of years. While believing living forms were introduced via special creation throughout these long days, they still allow for no large-scale evolution. The third faction of progressives agree that the days of Genesis are indeterminately long but differ from the other progressive factions in allowing some evolution—aided on various occasions by God's interventions of special creations of new higher order organisms—a sort of supernaturally punctuated evolution. The "gaps" in the fossil record are said to be accounted for by this explanation.

Theistic

Theistic evolutionists are also commonly referred to as "evolutionary creationists," or "providential evolutionists." Like the progressives, theists have no problem with old universe or the days in Genesis being indeterminately long. They think evolution occurred with no special interventions of special creation.

The theistic group has two main subgroups. The first subgroup holds that the randomness involved in evolution is minimal or nonexistent. So evolution is basically accepted with the proviso that the God of the Bible, not chance, decided the human outcome by directly guiding the process. These theists are usually referred to as *theistic evolutionists*. The second subgroup believes that evolution has an authentic random element, with God employing that randomness to produce the desired end of humans. The analogy that is often used to make this distinction is that of a casino. Many people playing the games see it from a perspective of randomness concerning whether they will win or lose on the next bet. Simultaneously, the casino owner views these games of chance as producing a predictable year-end profit. Most theists view the differences between the subgroups as a matter of design versus accident and not as creation versus evolution.

Thus far it sounds like very few biology instructors would have any problems with this type of theistic evolution. After all, whatever the ultimate causes may be, the theists appear to accept evolutionary processes. Unfortunately, that is not exactly the case with many theistic adherents. Problems arise for them when evolution is taught with an element of randomness that implies, as Stephen Jay Gould and others have written, that if the tape of life's history were rerun, then organisms other than humans would have evolved. Many theists take issue with that sort of evolution instruction.

Theists and progressives do not consider themselves to be creationists and generally do not enjoy being included in that categorization for two probable reasons: (1) Over the decades, literalists have been the ones recognized as creationists, and progressives and theists are not fond of being thought of as literalists. (2) Creationism has been scientifically discredited, and courts have ruled that it cannot be taught as science in public-school science classrooms. Therefore, those progressives and theists who want their version of evolution/creation taught in public-school science classrooms must distance themselves from the creationist label.

As for the ID group, they can be considered theists, and many do not mind that categorization. Some in the ID group, however, point out that their personal views are substantially different from theist views, because they avoid discussions of any pertinent theological positions they may hold. As mentioned previously, ID advocates do not like being called creationists even though some leading non-ID creationists think that ID advocates are creationists who camouflage their creationism as ID or progressive creationism. Whatever ID proponents claim, however, the overwhelming majority of people who study the anti-evolution phenomenon classify ID as creationism. The bottom line is that intelligent design theory purports that evolution instruction should be modified to include supernatural causation (or nonsupernatural intelligent design[s]) as an explanation in science courses instead of, or in addition to, evolutionary theory. There is a minute portion of ID advocates not classified as theistic because they claim to be agnostic or atheistic. Nevertheless, they are still considered creationists because they contend that an intelligent designer "created" and that this assertion should be included as science within evolutionary biology instruction.

In addition to encountering students having creationist ideas that were fostered outside of classroom instruction, biology instructors will encounter students having creationist ideas that were fostered by classroom instruction in religious private schools and home schools. In general, higher percentages of students graduating from religious private schools and home schools attend college than do students from public schools. Most students who attend private religious schools do not go on to religious colleges and universities but attend secular postsecondary institutions. Many of these students will have been taught about evolution in their high schools—some correctly and some incorrectly. A smaller number of these students will have been taught the strict creationist view of evolution being "bad science." Many of the students, and almost certainly the group just mentioned, will bring numerous misconceptions to college and university level biology courses that will need to be addressed if evolution is to be understood properly.

CONFRONTING VARIOUS EVOLUTION/CREATION ISSUES

This section includes a somewhat random collection of various issues that biology instructors may confront inside and outside of the classroom. Some students may bring these matters up, others may be held by students but never heard, while others may never be heard by a particular instructor but may be encountered by other instructors. Some of these issues may underpin misconceptions more directly related to understanding course content, while others may not. Whatever the case, a greater understanding of some of the conceptions that students hold relevant to understanding evolution is potentially helpful to a biology instructor. Each of the following ideas may be held by few or by many students, but the distinction will not be made in this section to reduce repetition such as "some students think . . ." "many students think . . ." and so forth. Moreover, the proportions are unknown much of the time.

"Gaps" in the Fossil Record

The matter of the fossil record lacking transitional forms or "missing links" is troubling for students. For whatever reason, students think that the fossil record is a somewhat perfect gradual progression of fossils, except that major important fossils are absent—especially between humans and what they consider to be apelike organisms. They think that these gaps are so monumentally wide that scientists are perplexed. They think that scientists really don't know enough to say for sure that humans evolved, but they do so because they are only making theories (i.e., guesses) anyway. Thus, students who have not thought much about evolution think that their ideas parallel those of scientists, with the exception that scientists conjecture that apelike organisms evolved to humans because it is their job to do so. They think that scientists know there is little-to-no evidence for evolution, but it is their best guess. Students often don't find such perceptions compelling.

Furthermore, antievolutionists use these supposed gaps extensively in their teachings against evolution. Thus, students who have encountered the teachings of creationist leaders almost invariably have heard something about the fossil record being inadequate to support evolution. Most high school and college students never take geology courses and never learn about paleontology. Therefore, it is helpful to students' understanding of evolution for biology instructors to briefly introduce the fossil evidence for evolution. Students need to understand why paleontologists are compelled by the evidence to conclude that evolution did occur. For example, provide students with some well-documented fossil records such as elephant lineages, and later introduce some hominid transitions (e.g., transitions from something rather apelike to modern humans over the last 3 million years: *Australopithecus africanus*, brain size ~450 ml; *Homo hablis*, ~750 ml; *Homo ergaster*, ~1,000 ml; *Homo sapiens*, ~1,350 ml.[16]

Dating fossils and rocks

Another topic helpful to students' understanding of evolution is how fossils and rocks are dated. As with their approach to gaps in the fossil record, antievolution leaders, particularly literalists, spend time attempting to demonstrate that the methods for dating fossils and rocks are inaccurate. The dating is so inaccurate, they contend, that there is no reason to consider the earth old enough for evolution to have taken place. Creationist leaders also attack sciences that lend support to evolution, and young earth creationists attack any science that illustrates an old earth, thus students may come to the biology classroom believing that most of astronomy, geology, and paleontology is not credible science. Whether or not students in a particular biology course have been exposed to such "teachings," they benefit from a brief overview on dating methods, which adds to their general evolution literacy.

Teaching "both sides"

Some creationist students and creationist leaders advocate the teaching of "both sides" in biology class. The general public is often taken in by this approach; after all, presenting the students with more generally sounds good, and presenting students various sides of an issue also sounds good. A typical example is the contention that the Second Law of Thermodynamics should be brought up in evolution instruction as counter evidence because, creationists contend, that an increasing complexity through evolution is a contradiction with this law. If instructors want to include the concept of why evolution does not contradict the Second Law for evolution education purposes, then that would certainly be beneficial to students. However, students and the public need to understand that teaching "both sides" or "various sides" is a discussion that exists only in the public arena and does not exist in the scientific community. There are no "sides," no controversy, to the factuality of evolution. In addition, creationism is not science. Articles on the supposed "creationism" side of evolution do not exist in any standard scientific journals because creationism is not science. No federal or state agency provides funding for scientific research or scientific conferences on creationism, because it is not science. There are no "creation science" degree programs at secular institutions. Although these facts are blatantly obvious to biology instructors, they may not be obvious to their students and to the general public. Creationism—in any form—does not represent "the other side" in science. Juxtaposing scientific concepts with religious beliefs and presenting non-science as science is not only bad for science understanding but also educationally unfair to students.

A threat to morality

It is curious to many instructors why some students and others feel so strongly that "correct" science (in their view) be taught. It is as if they care so much about science that they want some form of "science content police." Where does such a motivation come from?

For some, it is strictly engendered by religious motivations. Many students whether religious or not, think that accepting evolution as scientific fact somehow excuses people from moral responsibility. Thus, public acceptance of evolutionary theory necessarily leads to moral decay. In other words, if you teach people that they are descended from animals, then they will act like animals. These unfortunate characterizations have been widespread since Darwin's time. It seems students don't understand that there are large numbers of peoples of various cultures who accept evolution yet do not base their morals on nature and find no logic in doing so.

Many students who are religious perceive evolution as a threat to their morality as defined in the Bible. To them, the important point is not that the acceptance of evolution might result in people acting like animals but that it might undermine the moral authority of the Bible. To these students, there is a biblical foundation for the demarcation between right and wrong behavior. They believe that the Bible is inerrant and that evolution is incompatible with scripture. Teaching evolution, therefore, is an attempt to undermine biblical accuracy and attack scriptural integrity. Nevertheless, this way of thinking has more to do with theological disciplines than biological sciences. Some instructors have found it useful to point out that evolution doesn't necessarily lead to moral decay and that there are large numbers of people who find no problem with evolution and yet are upstanding religious citizens.

A THREAT TO SALVATION

The perceived attack on scriptural integrity goes further than morality issues. Many religiously literate evangelical Christians, for example, believe that they have a commandment given to them in each of the first four books, and Acts, of the New Testament. That commandment is to spread the gospel that Jesus was God on earth, died for everyone's sins, and rose from the dead. Everyone who believes this and repents of their sins, accepting Jesus as Lord of their life, will be Christians and go to heaven after death.[17] One of the most common questions in this regard for one Christian to ask someone else is: "Are you saved?" Nonbelievers becoming believers are known to have been "saved." Evangelists and preachers at the end of sermons will typically invite the unsaved in the audience to become saved today. Those not religiously inclined or exposed to this popular form of Christianity might ask from what are they being saved. They are being saved from going to hell after death. This is essentially what most evangelical Christians believe, and evangelical Christians appear to comprise most of the organized resistance against evolution education. In the U.S., evangelical Christians are more than 100 million strong—their power, influence, and ranks are swelling.[18]

Many creationists think that others exposed to evolution may come to believe that the Bible is inaccurate with regard to supernatural creation of humans, thus seeing no need to be beholden to the biblical Creator. Thus, evolution can be perceived as a significant barrier to the acceptance of Christianity. The world's leading creationist evangelist states: "I've found that evolution is one of the biggest, if not *the* biggest, stumbling block to people being receptive to the gospel of Jesus Christ."[19] Again, these ideas are religious and not scientific, but if college instructors discuss this notion with students during office hours, they might find it helpful to tell students that the overwhelming majority of Christian seminaries have little-to-no problem accepting both evolution and Christianity. Whether students or instructors care to engage in any of these primarily religious discussions, it is still helpful for biology instructors to understand the deep, personal, heartfelt reasons why students desire a de-emphasis of evolution or the inclusion of anti-evolutionary material in science courses. By understanding that some students are not necessarily being combative just for the sake of youthful rebelliousness, instructors can better help all involved.

WHY LEARN EVOLUTION?

Many instructors do an excellent job explaining to students why evolution is important to the life sciences and for general science understanding. Other instructors are so deeply involved over the years in the field of biology, that the explanation of why evolution is important seems blatantly obvious to them, and they assume it is obvious to students as well. Therefore, they spend relatively little time, if any, explaining this "obvious" importance to students. Some instructors just let the students read that part in the textbooks, while class time will be spent with the nitty-gritty of the real science. Yet others just assume that it was covered in high school biology or general science classes and is too rudimentary to take up precious college-level time. Nevertheless, there is pressure to deemphasize evolution in secondary schools in many parts of the nation, and some students, creationists, politicians, and administrators believe that students can attain a well-rounded science background without learning about evolution, so this topic is of vital importance in postsecondary education. College instructors should explain to students why evolution is important beyond indicating where we came from. Not only should biology instructors teach these reasons, but this topic should be included in student assessments as are other important aspects of biology. The following are some of those reasons student should know.

While much of biology explains *how* organisms function, evolution explains *why* organisms function as they do. It is the lens by which scientists in many fields interpret data. Life's unity and diversity have to do with organisms' relatedness and shared history via evolution. Why is form adapted to function? Why do organisms have a variety of non-adaptive features that coexist amidst those that are adaptive? Students should know that these are important questions and that only evolutionary theory can answer these *why* questions.

There are a host of subdisciplines of biology that are intertwined with evolution. Some of the more obvious are genetics and ecology. Others may not be so readily remembered by or obvious to students; for example, within systematics, cladistics classifies organisms based on their evolutionary history; without understanding evolution one cannot understand modern methods of classification. Within developmental biology, some embryological phenomena are only understood by evolutionary history. Outside of strict biology, the fields of geology and paleontology are interrelated by the chronology of the history of life on earth as preserved in fossils, and evolution helps explain those historical processes.

With regard to more contemporary issues close to home, the evolutionary consequences of incorrect antibiotic use by the American public has an immense impact on health and health care financing, costing more than $30 billion annually.[20] Approximately 14,000 people die each year from drug-resistant infections acquired in hospitals in the United States. In poor countries, the effects of antibiotic resistance are even more severe. These natural selection problems face everyone. Explanations of such phenomena and potential solutions are drawn from the science of evolution.

In what other ways does an understanding of evolution help in health care? Genetic diseases are better understood in an evolutionary context. Some diseases caused by complex interactions between genes and environmental factors are studied using evolutionary principles and approaches. And gene sequences that affect reproductive fitness are studied by molecular evolutionary biologists, who have developed methods to identify them.

Human and other animal behavior studies are enriched by evolutionary perspectives. Some areas of psychology related to how students learn are informed by evolution. For example, Howard Gardner, a professor of psychology at the Harvard Graduate School of Education concisely states why evolution is important in his field, again showing in importance of evolution in a field not readily obvious to many students:

> *"This is an important area of science, with particular significance for a developmental psychologist like me. Unless one has some understanding of the key notions of species, variation, natural selection, adaptation, and the like (and know how these have been discovered), unless one appreciates the perennial struggle among individuals (and populations) for*

survival in a particular ecological niche, one cannot understand the living world of which we are part."[21]

Students should learn that not only is evolution present in psychology, it permeates other fields such as history, philosophy, literature, and the arts. In its unifying nature, evolutionary theory is the overarching principle and underlying theme of the biological sciences, underpinning and permeating all life sciences. It is a context that brings together the morphological, physiological, behavioral, or biochemical characteristics of living things. It helps explain how the organisms of today got to be the way they are, how they are related and, to some extent, why they behave the way they do. Without evolutionary theory, biology would be unconnected, disparate facts with no scientific answers to the *why* questions. The sciences based on evolution would no longer be helpful to science or society.

If students do not learn evolution, they will not comprehend evolutionary connections to the other scientific fields, not fully understand the living world, nor be fully biologically literate. The field of biology cannot be fully understood without some understanding of evolution and students need to be taught these types of reasons why evolution is important to know, value, and assess as other important aspects of the course. If students have not learned some good reasons why evolution is vital to the biological sciences by the end of an introductory biology course, they are missing something that most instructors would probably say is of great importance. Instructors who find this is the case at the end of their courses should change their teaching methods so that students will understand why the evolution they are learning is important to know.

ACADEMIC ACTION

In the last 15 years the higher education academic community has increased its efforts in the area teaching and learning evolution. Some examples follow, but much more than this has been accomplished. There have been multiple national evolution education conferences concerning research, pedagogy, creation of materials, dissemination, and activism. Education committees have been formed in professional societies such as the Society for the Study of Evolution (SSE). Such societies have issued statements and materials supporting evolution education, with eight societies joining together to author a white paper on the science of evolution.[22] This paper is useful for policymakers, educators, and scientists. Numerous societies have helped give sessions on evolution for high school science teacher conferences at the local, state, national, and international levels. Recently, with NSF support, 41 professional society representatives met with those of the National Center for Science Education (NCSE) and produced, among other things, a resource matrix—with numerous recommended resource sites—and a blueprint for societies on how to construct evolution education workshops for teachers (both located at: http://www.ucmp.berkeley.edu/ncte). McGill and Harvard University personnel have created an evolution education research center. Numerous academics have authored books and articles, and have given media interviews on teaching and learning evolution.

One of many things that can be done at the college or university classroom level is to help future high school biology teachers. Although some students know they will become biology teachers, others remain undeclared when they take their first post-secondary biology course. Some students will return to university after switching careers for the sole purpose of becoming biology teachers. In any case, biology instructors may want to address the problems of teaching evolution in public schools to help satisfy some students' immediate and possibly future needs. After all, a recent state-by-state evaluation of the treatment of evolution in science standards determined that "more than one-third of all states do not do a satisfactory job." Astonishingly, "ten [states] never use the 'E-word.'"[23] By helping future biology teachers at the college level, instructors are indirectly helping students who may never take a college-level biology course. For most students the last formal science course they take is high school biology.

Once future teachers become practicing teachers in the schools, they may often encounter pressure to deemphasize evolution in their own public high school biology classes. There are ways to help resist this pressure, and one of the most powerful is knowledge of court decisions regarding teaching evolution in public schools. Unfortunately, many high school teachers have a poor understanding of these court decisions, which could help them defend teaching evolution. The long-time editor of the *American Biology Teacher*, a university biology professor, recommends that these court decisions be incorporated into college-level biology courses (and possibly others) so that future teachers will be better prepared. (For more details see *BioScience*, September 2004, and the Web site of the National Center for Science Education.)[24] Public high school biology teachers not only need to understand biology but also need to understand how to defend their teaching of it.

With regard to the public rejecting evolution, some instructors think that their teaching about evolution will not make much of a difference. We contend that evolution education is necessary and productive. National polls correlating education and agreement with evolution are inconclusive by their very nature. Nevertheless, polls conducted by the Gallup organization indicate that as the level of respondents' general education increases, the level of agreement with the idea that God created humans basically in their present form within the last 10,000 years decreases. For example, about 55% of respondents with a high school education or less agreed with this idea, while approximately 45% with some college education agreed. About 40% of college graduates agreed, and only about 30% of postgraduates agreed.[25]

Similarly, a poll conducted by the National Science Board (NSB) reports that as general education increases so does agreement with humans developing from earlier species of animals. For example, those respondents with less than a high school education were least apt to agree (~30%), those having graduated from high school were next in agreement (~40%), those with a baccalaureate degree followed next (~60%), and those with graduate or professional degrees had the highest proportion of respondents in agreement (~75%). The NSB reported that agreement also increased with educational levels in science education. For example, those with the least amount of science education agreed least (~35%), those in the middle category of science education were next in agreement (~50%), and those with a high amount of science education were most in agreement (~65%).[26] The direction of causality is not evident; while it may appear that education influences agreement, it may be the case that those who are in agreement are more likely to further in their education, while those who are not in agreement are less likely to further their education. There are many confounding factors that could come into play regarding these types of polls, but they are reported here for those interested in seeing poll results.

One confounding issue is simply the word "evolution;" people often do not know what the word means. And even when the rudimentary meaning of this word is somewhat understood, some problems stem from confusion of other scientific words with everyday words. Research results indicate that non-majors confuse the scientific terms "adapt," "adaptation," and "fitness" with the everyday terms they use outside of science. Students who are accustomed to the word "adapt" referring to individuals changing their behavior in response to environmental factors sometimes have problems learning the evolutionary meaning. For example, students who think of someone who has adapted to staying up late for a part-time job often transfers that meaning to evolutionary adaptation and thinks that the environment acts on individual organisms forcing them to change their characteristics or perish. Likewise, students think that an athletic star's fitness is better than the fitness of others' because of their ability to perform superior physical feats. Students may also think of fitness in the sense that some students are more fit for medical school than are others, and they may apply these meanings in an evolutionary context.

Students also misunderstand the meanings of more general scientific terms such as "theory" and "law." A large number of students think that theories are far inferior to laws in the scientific sense. Some believe that if theories are good enough and have enough evidence, that theories will become laws. Outside of science, students have overwhelmingly heard the word "theory" to mean something with very little-to-any supporting evidence. In either of these cases, inside or outside of science, experiences of numerous students with the word "theory" has been in the sense of: "It's *only* a theory."

Not only are students' misconceptions often reinforced by vernacular meanings of terminology, but it is also difficult for many students to change their conceptions. Students may appear to change their conceptions of the scientific meanings of these words if they are allowed to regurgitate definitions on knowledge-level assessments. However, if these students were to be assessed one year later, many would have reverted to the meaning they have known most of their lives. The reason is similar to that discussed previously in the misconceptions section.

It is often surprising how inappropriate use of the word "theory" can infiltrate even persons with advanced degrees in the life sciences and related fields. One example (and there are many) comes from a meeting we had with a Boston physician, who has for years practiced medicine, conducted medical research, and taught in medical schools. In conversation over a meal, the problems of evolution education arose and how the general public does not agree much with the accuracy of evolutionary science. After about 10 minutes explaining the current state of affairs in evolution education to the physician, he made clear he certainly accepted evolutionary theory as being a science with all its great rigor and evidence to support it. He also made clear that he thinks the diversity of life is accountable to evolutionary processes, and that it's logical and scientific to come to that conclusion.

In spite of this, he continued to say, he understands why others don't feel as strongly as he does, because evolution is not as credible as other sciences because it happened in the distant past and no one was present to collect data. Therefore, he continued, that is why evolution is *only* a theory—and it will always be *only* a theory. This physician had no religious conflicts with evolution, and no scientific or pseudoscientific rationales against evolution, yet felt that the "theory" label is appropriately attached to indicate evolution's close proximity to the bottom rungs of science. With semantics alone, biology instructors have their work cut out for them.

ORIGIN OF LIFE

When the origin of life is discussed in college biology courses, there will probably be some students with perceived religious conflicts. For this reason, many instructors do not include teaching the origin of life in their biology courses, often saying that concerns prebiotic evolution and is not appropriate for a biology course. Others contend that prebiotic evolution is too abstract and complex for the introductory level courses, or that this topic belongs in a chemistry or geophysics course. On the other side of this fence are instructors who think as Theodosius Dobzhansky proclaimed, that "evolution is a process which has produced life from non-life, which has brought forth man from an animal, and which may conceivably continue doing remarkable things in the future."[27] They think the biology should include the origination of life. Other instructors advocate inclusion as an important aspect of college students' scientific literacy.

When the origin of life is included in a biology course, some students think that their religious beliefs are challenged despite the fact that most religious leaders accept that God may have chosen to use evolution (including prebiotic evolution) to create the diversity of life on earth. This is a problem of religious illiteracy. Many students just assume that evolution is not accepted within their religion. Because of the situation, some biology instructors distribute to their students official statements from major religious organizations that support evolution. Both creationist and non-creationist students are often surprised to learn that most religious leaders find evolution to be compatible with their faiths. This helps lessen the common perception of science being in conflict with all or most religions. It becomes clearer to students that it is really just a dispute with a small number of religious factions. Such action has helped diminish student fear of the subject, improved student attitudes about learning it, and lessened student animosity toward the instructor.

In addition to perceived religious reasons, many creationists have non-religious rationales for their rejection of any prebiotic evolutionary science. The arguments are standard fare in the creationist literature and basically have to do with the unknown biochemical pathways/mechanisms, mathematical improbabilities, thermodynamic impossibilities (perceived increasing complexity instead of entropy), and irreducible complexity (a term used by ID advocates), which basically posits that at early evolutionary stages the complexity is too low to have any function. Whether students are creationists or not, these matters could underpin some misunderstandings about prebiotic evolutionary science and may be worth considering in most introductory courses.

PHILOSOPHY AND SCIENCE

One of the major stumbling blocks many creationist students have with the science of evolution is understanding the philosophical underpinnings of science. One of the most important scientific underpinnings for helping students in this regard is understanding the role methodological naturalism (MN) has in science, and how that role does not conflict with supernaturalism. While the following may be almost common sense to science instructors, many students have not been taught and/or have not learned what those instructors assume they know. Students need to understand that when scientists do their scientific work they are applying MN, thus looking only for natural causes and not attributing phenomena to supernatural causation.

Some students wonder how individuals can apply naturalism to their work, while believing in the supernatural. We have found in our limited personal experience that the following example seems to help students—although often not immediately; students need to process this information for awhile, possibly months. The example go something like this: When a person who believes in the supernatural goes to get their car repaired because something in the engine is malfunctioning, they want the mechanic to look for, and hopefully find, a natural cause. The car owner certainly does not mind if the mechanic prays to a supernatural being asking for assistance before, during, and after the diagnosis process. However the car owner does not want the mechanic to report that the engine malfunctioned due to supernatural causes. The car owner wants the mechanic to report the natural cause for the malfunction and what natural remedy will correct the malfunction. So while car mechanics may be devoutly religious, they apply methodological naturalism as standard practice—the rule—of auto mechanics. If auto mechanics were to report to car owners that malfunctions were due to supernatural causes and that nothing could be done to remedy the situation, even devoutly religious clients would most likely take their cars to other mechanics for diagnosis.

This example must be used with the greatest of care because it is very easy to offend in this regard. Devoutly religious students sometimes think that it is demeaning to say a car engine is malfunctioning due to God's actions. That is why it is best to attribute it to "supernatural" or "extraterrestrial" causes. The demeaning feeling often stems from the assumption that God would not do something as trivial as causing an engine's malfunction or that someone would attribute such relatively unimportant phenomenon to God. The fear of many leading creationists and some students is that MN will be used to examine religion. Students should come to understand that MN cannot judge possible supernatural existence or phenomena. A little evolutionary biology can stir powerful feelings among students.

Endnotes

Chapter 1: Educational Research and Improving Teaching

1. National Research Council Committee on Undergraduate Science Education 1997, p. v
2. Redish, 2003, p. 15
3. National Research Council Committee on Scientific Principles for Education Research, 2002, p. 1
4. National Research Council Committee on Scientific Principles for Education Research,2002, pp. 3–5
5. National Research Council Committee on Scientific Principles for Education Research, 2002, p. 49
6. National Research Council Committee on Undergraduate Biology Education to Prepare Research Scientists for the 21st Century, 2003, p. 7
7. Nelson, 2004
8. National Research Council Committee on Undergraduate Biology Education to Prepare Research Scientists for the 21st Century, 2003, p. 66
9. National Research Council Committee on Undergraduate Biology Education to Prepare Research Scientists for the 21st Century, 2003, p. 87
10. National Research Council Committee on Undergraduate Biology Education to Prepare Research Scientists for the 21st Century, 2003, p. 91
11. Jenkins et al, 2003, p. 61
12. Many of the principles/properties of effective teaching noted in this chapter are based on Ramsden, 2003.
13. Jenkins et al, 2003, p. 68
14. Ramsden, 2003, pp. 86–87
15. Bligh, 2000
16. National Research Council Committee on Undergraduate Biology Education to Prepare Research Scientists for the 21st Century, 2003, p. 3
17. Uno, 2002, p. 2

18. National Research Council Committee on Undergraduate Biology Education to Prepare Research Scientists for the 21st Century, 2003, p. 27–48.
19. BSCS, 1994; Uno 2002, p. 117
20. Society for College Science Teachers, 1993, p. 31
21. Nelson, 2004, p. 132
22. Nelson, 2004, p. 137

Chapter 2: Teaching & Learning

1. Ramsden, 2003, p. 88
2. Biggs, 1999, p. 4
3. Biggs, 1999, p. 4
4. Bloom, 1956
5. Adapted from BSCS 1994, Uno 2002, p. 76
6. National Science Board, 2000
7. Schneps & Sadler, 1988, 1997
8. Adapted from Gardiner, 1998
9. Hake, 1998
10. Sadler, 1998
11. Tobin et al., 1994, p. 47
12. Lawson, 1994, p. 166
13. Lawson & Weser, 1990
14. Alters & Nelson, 2002, p.1896
15. Springer et al., 1997; Hake, 1998; Gardiner, 1998
16. McKeachie, 1994; Hake, 1998; Springer et al., 1997; Gardiner, 1998; Nelson, 1994
17. Sundberg & Dini, 1993
18. National Research Council Committee on Undergraduate Biology Education to Prepare Research Scientists for the 21st Century, 2003, p. 72
19. National Research Council Committee on Undergraduate Biology Education to Prepare Research Scientists for the 21st Century, 2003, p. 60
20. http://www.bscs.org/page.asp?id=curriculum_development|key_features_of_our_programs|BSCS_5Es [accessed October 27, 2004]
21. Uno, 2002, p. 54
22. D'Avanzo & McNeal, 1996
23. National Research Council Committee on Undergraduate Biology Education to Prepare Research Scientists for the 21st Century, 2003
24. National Research Council Committee on Undergraduate Biology Education to Prepare Research Scientists for the 21st Century, 2003, pp. 77, 81

25. National Research Council Committee on Undergraduate Biology Education to Prepare Research Scientists for the 21st Century, 2003, pp. 81–82
26. National Research Council Committee on Undergraduate Biology Education to Prepare Research Scientists for the 21st Century, 2003, p. 84
27. National Research Council Committee on Undergraduate Biology Education to Prepare Research Scientists for the 21st Century, 2003, p. 84
28. National Research Council Committee on Undergraduate Biology Education to Prepare Research Scientists for the 21st Century, 2003, p. 85
29. Adapted from National Research Council Committee on Undergraduate Biology Education to Prepare Research Scientists for the 21st Century, 2003, pp. 85–86
30. Bransford et al., 2000, pp. 14–15
31. Committee on Undergraduate Science Education, 1997, p. 28
32. Mayr, 1982
33. Committee on Undergraduate Science Education, 1997; Alters & Nelson, 2002, p. 1894–1895
34. Bishop & Anderson, 1986; Bishop & Anderson, 1990; Jensen and Finley 1996; Alters & Alters, 2001.
35. Greene, 1990, p. 883
36. Trowbridge & Wandersee, 1994
37. Weimer, 2002
38. Bligh, 2000, p. 4
39. Uno 2002, p. 99
40. Gardner, 1999b, pp. 33–34
41. Bligh, 2000, p. 205
42. Jenkins et al., 2003, p. 43–44
43. Bligh, 2000
44. Bligh, 2000
45. Bligh, 2000
46. Bligh, 2000

Chapter 3: Assessment

1. Uno, 2002 p.87
2. Jenkins et al., 2003, p.69
3. BSCS, 1994 ; Uno, 2002, p.47
4. Sadler, 1998
5. Ramsden, 2003
6. Ramsden, 2003
7. Ramsden, 2003

8. Uno, 2002, p.3
9. Alters & Alters, 2001

Chapter 4: Evolution Vs. Creationism

1. Pinker, 2002, p. 223
2. Gallup Poll News Service, 2004
3. National Science Board, 2000
4. For additional data and specific data concerning this polling, see Alters & Alters, 2001; Alters & Nelson, 2002.
5. National Science Board, 2000
6. People for the American Way Foundation, 2000
7. Lawson and Weser, 1990; Sinclair et al., 1997; Sinclair & Pendarvis, 1998; Dagher & BouJaoude 1997; Brickhouse et al., 2000
8. Institute for Creation Research, n. d.
9. Alters & Alters, 2001, p. 53; and Scott, 2004
10. See Chapter 4.
11. McInerney 1997
12. Numbers, 1998; Tidon & Lewontin, 2004
13. Stephen Jay Gould in Alters & Alters, 2001, p. 4
14. Forrest & Gross, 2004
15. Forrest & Gross, 2004, p.8
16. Eldredge, 2000
17. There are many variations of this oversimplification of the gospel. The purpose here is not to elaborate on all the denominational nuances but merely to give the basics to illustrate the point. Almost all of the hundreds of creationists with whom the authors have discussed this matter have this basic understanding.
18. CNN, 2004
19. Ham, 1998, p. 73
20. Palumbi, 2001
21. Gardner, 1999a, p.16
22. Futuyma, (Ed.), 2000
23. Lerner, 2000, p.xii
24. Moore, 2004
25. The Gallup Organization, Inc. Survey dates: August 24 – 26, 1999.
26. National Science Board, 1996
27. Dobzhansky, 1967, p. 409

References

Alters, B. J. & Alters, S. M. (2001). *Defending evolution in the classroom: A guide to the creation/evolution controversy*. Sudbury, MA: Jones & Bartlett.

Alters, B. J. & Nelson, C. E. (2002). Perspective: Teaching evolution in higher education. *Evolution, 56,* 1891–1901.

Biggs, J. (1999). *Teaching for Quality Learning at University*. Buckingham, England: SRHE and Open University Press.

Biological Sciences Curriculum Study (BSCS). (1994). *Developing biological literacy*. Dubuque, IA: Kendall/Hunt.

Bishop, B. A., & Anderson, C. W. (1986). *Evolution by natural selection: A teaching module*. (Occasional Paper No. 91). East Lansing: Institute for Research on Teaching, Michigan State University. Available from the Education Resources Information Center (ERIC). ED 272 383.

Bishop, B. A., & Anderson, C. W. (1990). Student conceptions of natural selection and its role in evolution. *Journal of Research in Science Teaching, 27,* 415–427.

Bligh, D. A. (2000). *What's the use of lectures?* San Francisco: Jossey-Bass.

Bloom, B. S. (Ed.). (1956). *Taxonomy of educational objectives*. New York: David McKay.

Bransford, J. D., Brown, A. L., & Cocking, R. R. (2000). *How people learn*. Washington, DC: National Academy Press.

Brickhouse, N. W., Dagher, Z. R., Letts, W. J., & Shipman, H. L. (2000). Diversity of students' views about evidence, theory, and the interface between science and religion in astronomy course. *Journal of Research in Science Teaching, 37,* 340–362.

CNN. (2004, October 30). *CNN presents: The fight over faith* [Television broadcast].

Committee on Undergraduate Science Education. (1997). *Science teaching reconsidered: A handbook*. Washington, DC: National Academy Press.

Dagher, Z. R., & BouJaoude, S. (1997). Scientific views and religious beliefs of college students: The case of biological evolution. *Journal of Research in Science Teaching, 34*, 429–445.

D'Avanzo, C. & McNeal, A. (1996). Inquiry teaching in two freshman level courses: Same core principles by different approaches. In C. D'Avanzo & A. McNeal (Eds.), *Student-active science: Models of innovation in college science teaching*. Philadelphia: Saunders.

Dobzhansky, T. (1967). Changing man: Modern evolutionary biology justifies an optimistic view of man's biological future. *Science, 155*, 409–415.

Eldredge, N. (2000). *The triumph of evolution and the failure of creationism*. New York: Freeman.

Forrest, B. & Gross, P. R. (2004) *Creationism's trojan horse: The wedge of intelligent design*. New York: Oxford University Press.

Futuyma, D. J. (Ed.). (2000). *Evolution, science and society: Evolutionary biology and the national research agenda*. Piscataway, NJ: Office of University Publications, Rutgers.

Gallup Poll News Service. (2004, November 19). Third of Americans say evidence has supported Darwin's evolutionalry theory. Princeton, NJ: The Gallup Organization, Inc.

Gardiner, L. F. (1998). Why we must change: The research evidence. *Thought and Action, 14,* 71–88.

Gardner, H. (1999a). *The disciplined mind: What all students should understand.* New York: Simon & Schuster

Gardner, H. (1999b). *Intelligence reframed: Multiple intelligences for the 21st century*. New York: Basic Books.

Green, E. D. (1990). The logic of university students' misunderstanding of natural selection. *Journal of Research in Science Teaching, 27,* 875–885.

Hake, R. R. (1998). Interactive-engagement versus traditional methods: A six-thousand-student survey of mechanics test data for introductory physics courses. *American Journal of Physics, 66,* 64–74.

Ham, K. (1998). *Creation evangelism for the new millennium*. Green Forest, AR: Master Books, Inc.

Institute for Creation Research. (n.d.). *ICR Tenets of Creationism*. Retrieved October 31, 2004 from http://www.icr.org/abouticr/tenets.htm.

Jenkins, A., Breen, R., Lindsay, R., & Brew, A. (2003). *Reshaping teaching in higher education*. London: Kogan Page Limited.

Jensen, M. S. & Finley, F. N. (1996). Changes in students' understanding of evolution resulting from different curricular and instructional strategies. *Journal of Research in Science Teaching, 33,* 879–900.

Lawson, A. E. (1994). Research on the acquisition of science knowledge: Epistemological foundation of congnition. In D. L. Gabel (Ed.), *Handbook of research on science teaching and learning* (pp. 131–176). New York: Macmillan.

Lawson, A. E., & Weser, J. (1990). The rejection of nonscientific beliefs about life: Effects of instruction and reasoning skills. *Journal of Research in Science Teaching, 27,* 589–606.

Lerner, L. S. (2000). *Good science, bad science: Teaching evolution in the states.* Washington, DC: Thomas B. Fordham Foundation.

McInerney, J. (1997). Evolution of the NABT statement on the teaching of evolution. *Reports of the National Center for Science Education, 10* (1), 30–31.

McKeachie, W. J. (1994). *Teaching tips: Strategies, research, and theory for college and university teachers.* Lexington, MA: Heath.

Moore, R. (2004). How well do biology teachers understand the legal issues associated with the teaching of evolution? *BioScience, 54*, 860–865.

National Research Council Committee on Scientific Principles for Education Research. (2002). *Scientific research in education.* Washington, DC: National Academy Press.

National Research Council Committee on Undergraduate Biology Education to Prepare Research Scientists for the 21st Century. (2003). Bio 2010: *Transforming undergraduate education for future research biologists.* Washington, DC: National Academy Press.

National Research Council Committee on Undergraduate Science Education. (1997). *Science teaching reconsidered: A handbook.* Washington, DC: National Academy Press.

National Science Board (1996). *Science and engineering indicators* (NSB 96–21). Washington, DC: U.S. Government Printing Office.

National Science Board (2000). *Science and engineering indicators* (NSB-00–1). Washington, DC: U.S. Government Printing Office.

Nelson, C. E. (2004). The research-teaching-research cycle. In W. E. Becker & M. L. Andrews (Eds.), *The scholarship of teaching and learning in higher education: Contributions of research universities* (pp. 129–142). Indianapolis: Indiana University Press.

Numbers, R. L. (1998). *Darwinism comes to America.* Cambridge, MA: Harvard University Press.

Palumbi, S. R. (2001). *The evolution explosion: How humans cause rapid evolutionary change.* New York: Norton.

People for the American Way Foundation. (2000, March). *Evolution and Creationism in Public Education: An In-depth Reading of Public Opinion* Retrieved October 31, 2004 from http://www.pfaw.org/pfaw/general/default.aspx?oId=2095

Pinker, S. (2002). *The blank slate: The modern denial of human nature.* New York: Viking.

Ramsden, P. (2003). *Learning to teach in higher education* (2nd ed.). London: RoutledgeFalmer.

Rechler, S. A. (1999, November 9). Scholars debate creationist theories. *Harvard Crimson, 1*, 3.

Redish, E. F. (2003). *Teaching Physics with the Physics Suite.* Hoboken, NJ: Wiley.

Sadler, P. M. (1998). Psychometric models of student conceptions in science; Reconciling qualitative studies and distractor-driven assessment instruments. *Journal of Research in Science Teaching, 35,* 265–296.

Schneps, M. H. & Sadler, P. M. (1988). *A private universe*. Santa Monica, CA: Pyramid Films.

Schneps, M. H. & Sadler, P. M. (1997). *Minds of our own: Can we believe our eyes?* Cambridge, MA: Harvard-Smithsonian Center for Astrophysics.

Sinclair, A., & Pendarvis, M. P. (1998). Evolution versus conservative religious beliefs: Can biology instructors assist students with their dilemma? *Journal of College Science Teaching, 27*, 167–170.

Sinclair, A., Pendarvis, M. P., & Baldwin, B. (1997). The relationship between college zoology students' beliefs about evolutionary theory and religion. *Journal of Research and Development in Education. 30*, 118–125.

Society for College Science Teachers. (1993). A society for college science teachers position statement on introductory college-level science courses. *Journal of College Science Teaching, 23,* 31.

Springer, L., Stanne, M. E., & Donovan, S. (1997). Effects of small-group learning on undergraduates in science, mathematics, engineering and technology, a meta-analysis. Retrieved October 27, 2004, from http://www.wcer.wisc.edu/nise/CL1/CL/resource/scismet.htm

Sundberg M. D. & Dini, M. L. (1993). Science majors versus nonmajors: Is there a difference? *Journal of College Science Teaching, 23,* 299–304.

Tidon, R. & Lewontin, R. C. (2004). Teaching evolutionary biology. *Genetics and Molecular Biology, 27*(1), 124–131.

Tobin, K., Tippins, D. J., & Gallard, A. J. (1994). Research on instructional strategies for teaching science. In D. L. Gabel (Ed.), *Handbook of research on science teaching and learning* (pp. 45–93). New York: Macmillan.

Trowbridge, J. E., & Wandersee, J. H. (1994). Identifying critical junctures in learning in a college course on evolution. *Journal of Research in Science Teaching, 31,* 459–473.

Uno, G. E. (2002). *Handbook on teaching undergraduate science courses: A survival training manual.* Pacific Grove, CA: Brooks Cole.

Weimer, M. (2002). *Learner-centered teaching: Five key changes to practice*. San Francisco: Jossey-Bass.

Index